Guide for the Total Solar Eclipse of 2027

Color Edition

Fred Espenak

Edition 1.0
July 2024

Guide for the Total Solar Eclipse of 2027 – Color Edition

Astropixels Publishing
P.O. Box 16197
Portal, AZ 85632

Astropixels Publishing Website: *www.astropixels.com/pubs*

This book may be ordered at: *www.astropixels.com/pubs/Guide2027.html*

Printed in the United States of America

ISBN 978-1-941983-47-8

Astropixels Publication: AP045 (Version 1.0b)

First Edition

Front Cover: 1) A map of the path of the 2027 total solar eclipse through Africa and the Arabian Peninsula.
2) Background: Base relief of Egyptian Goddess Isis in the Tomb of Seti I, Abydos, Egypt. ©2006, F. Espenak.

Back Cover: 1) Tomb of Seti I, Abydos, Egypt, ©2006, F. Espenak.
2) Fred Espenak in Temple of Amon-Ra: Copyright ©1995 by F. Espenak.

More about the eclipse can be found at:

www.eclipsewise.com/solar/SEnews/TSE2027.html

Table of Contents

Photo 1 – The The Diamond Ring effect is seen just before totality begins.
Total solar eclipse of July 2, 2019 from Vicuna, Chile. © 2019 F. Espenak, www.MrEclipse.com

Photo 2 – The Sun's corona is only visible to the naked eye for a few minutes during a total eclipse of the Sun.
Total solar eclipse of August 21, 2017 from Casper, Wyoming. © 2017 F. Espenak, www.MrEclipse.com

Figure 1 – The path of totality of the Total Solar Eclipse of Aug 2, 2027 crosses Africa and the Arabian Peninsula.

1 – Introduction

On Monday, August 2, 2027, a total eclipse of the Sun will be visible from the Eastern Hemisphere. Although a partial eclipse will be seen from much of Europe, Africa, and South Asia the total phase in which the Moon completely covers the Sun (known as totality) will only be seen from within the narrow path of the Moon's umbral shadow as it sweeps cross ten countries: Spain, Morocco, Algeria, Tunisia, Libya, Egypt, Sudan, Saudi Arabia, Yemen and Somalia.

During totality, the Sun's faint and exquisitely beautiful outer atmosphere – the solar corona – can be seen with the naked eye, the landscape is plunged into an eerie twilight, and brighter stars and planets appear. A total solar eclipse is one of the most awesome natural spectacles, but to witness it you must be in the *path of totality*.

The path of the Moon's umbral shadow begins in the North Atlantic and travels east to reach southern Spain, Morocco and Gibraltar. The central line duration of totality here is a generous 4 minutes 52 seconds – exceeding the maximum duration of the 2024 total eclipse. The track skirts the Mediterranean coast of Africa as the umbra races across Algeria, Tunisia, and Libya.

The trajectory begins curving southeast as it crosses central Egypt and encounters Luxor, site of the Ancient Egyptian city of Thebes. Here lie some of the world's greatest archeological treasures including the Great Temple of Karnak and (across the Nile) the Valley of the Kings. Nearby, totality reaches its maximum duration of 6 minutes 23.2 seconds. No other total eclipse exceeds this duration for the remainder of the 21st Century.

The southern edge of the path includes a small corner of Sudan before crossing the Red Sea. Western Saudi Arabia falls within the path including the Islamic cities of Jeddah and Makkah (Mecca). After crossing western Yemen, the path sweeps over the Gulf of Aden before making final landfall in northeastern Somalia.

The last 44 minutes of the total eclipse brings the Moon's umbra across the Indian Ocean where it ends southeast of India.

A partial eclipse will be visible from a much larger region including most of Europe, north Africa, central and south Asia. Eclipse glasses or some type of eye protection must be used for viewing the partial eclipse. Of course, those in the path of the total eclipse must remove protective glasses to witness totality itself.

This guide is designed to assist observers in locating and reaching the central eclipse path to enjoy the awesome experience of totality.

Total Solar Eclipse of 2027 Aug 02

Greatest Eclipse = 10:07:50.2 TD (= 10:06:41.0 UTC)

Eclipse Magnitude = 1.0790
Gamma = 0.1421

Saros Series = 136
Saros Member = 38 of 71

Sun at Greatest Eclipse
(Geocentric Coordinates)
R.A. = 08h49m26.9s
Dec. = +17°45'41.3"
S.D. = 00°15'45.5"
H.P. = 00°00'08.7"

Moon at Greatest Eclipse
(Geocentric Coordinates)
R.A. = 08h49m40.1s
Dec. = +17°53'47.8"
S.D. = 00°16'43.1"
H.P. = 01°01'21.4"

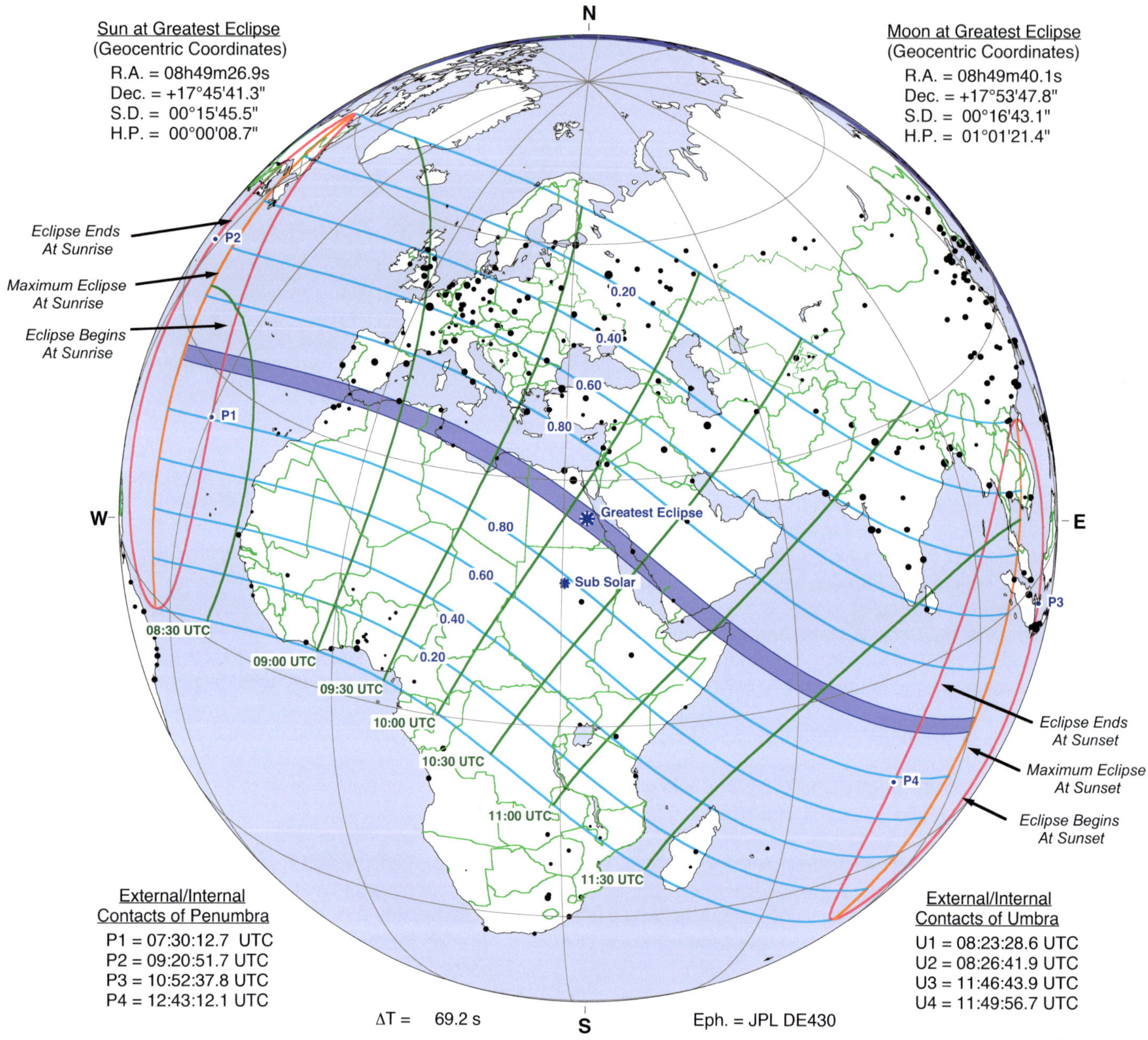

External/Internal
Contacts of Penumbra
P1 = 07:30:12.7 UTC
P2 = 09:20:51.7 UTC
P3 = 10:52:37.8 UTC
P4 = 12:43:12.1 UTC

External/Internal
Contacts of Umbra
U1 = 08:23:28.6 UTC
U2 = 08:26:41.9 UTC
U3 = 11:46:43.9 UTC
U4 = 11:49:56.7 UTC

ΔT = 69.2 s

Eph. = JPL DE430

Circumstances at Greatest Eclipse: 10:06:41.0 UTC

Lat. = 25°30.2'N	Sun Alt. = 81.7°
Long. = 033°11.0'E	Sun Azm. = 202.0°
Path Width = 257.7 km	Duration = 06m22.6s

Circumstances at Greatest Duration: 10:00:24.6 UTC

Lat. = 26°48.9'N	Sun Alt. = 80.9°
Long. = 031°07.9'E	Sun Azm. = 177.8°
Path Width = 257.2 km	Duration = 06m23.2s

0 1000 2000 3000 4000 5000
Kilometers

©2024 F. Espenak
www.EclipseWise.com

Figure 2 – Orthographic projection map of the Total Solar Eclipse of 2027.
Based on map from ***21st Century Canon of Solar Eclipses*** *by Fred Espenak (2017)*

2 – Orthographic Projection Map of the 2027 Total Solar Eclipse Path

The orthographic projection map of Earth in *Figure 2* shows the path of penumbral (partial) and umbral (total) shadows of the Moon during the eclipse.

The limits of the Moon's penumbral shadow define the region of visibility of the partial eclipse. It covers about half of Earth's daylight hemisphere and consists of several distinct zones or limits. At the northern and southern boundaries lie the limits of the penumbra's path. Great loops at the western and eastern extremes of the penumbra's path identify the areas where the eclipse begins and ends at sunrise and sunset, respectively.

Bisecting the *eclipse begins and ends at sunrise and sunset* loops is the curve of maximum eclipse at sunrise (western loop) and sunset (eastern loop). Points *P1* and *P4* mark the coordinates where the penumbral shadow first contacts (partial eclipse begins) and last contacts (partial eclipse ends) Earth's surface. The path of the umbral shadow (*path of totality*) appears within the penumbral path. Curves of constant eclipse magnitude define points where the maximum magnitude has values of 0.2, 0.4, 0.6, and 0.8.

Above the map is the instant of greatest eclipse[1], in Terrestrial Time[2] (TT) and Coordinated Universal Time[3] (UTC). The *Eclipse Magnitude* is the geocentric ratio of diameters of the Moon and the Sun. *Gamma* is the minimum distance of the Moon's shadow axis from Earth's center in Earth radii at greatest eclipse. Finally, the *Saros Series* of the eclipse is listed, followed by the *Saros Member* that gives the sequence position of the eclipse in the Saros, and the total number of eclipses in the series

In the upper left and right corners are the geocentric coordinates of the Sun and the Moon, respectively, at the instant of greatest eclipse. They are:

R.A. – Right Ascension
Dec. – Declination
S.D. – Apparent Semi-Diameter
H.P. – Horizontal Parallax

To the lower left are the contact times of the Moon's penumbral shadow with Earth. They mark the times when the partial eclipse begins and ends, globally. To the lower right are the contact times of the Moon's umbral shadow with Earth. They mark the times when the total eclipse path begins and ends, globally.

At bottom left are the geographic coordinates of the position of greatest eclipse along with the local circumstances at that location (i.e., Sun altitude, Sun azimuth, path width and duration of totality).

At bottom right are the corresponding values for the position of greatest duration[4]. Although the geographic locations of greatest duration and greatest eclipse differ by about 256 km (159 miles) along the eclipse track, the difference in the duration of totality is only 0.60 second.

3 – Overview Map of the Path of Totality

The overview map in *Figure 3* focuses on the path of the total eclipse through North Africa and the Arabian Peninsula. The white lines running across the eclipse path mark the position of mid-eclipse at 5-minute intervals. The positions of *Greatest Eclipse* and *Greatest Duration* (both in Egypt) are marked *GE* and *GD*, respectively.

The Mercator projection used in this map preserves local directions and shapes at the expense of inflating the size of objects the further they are from the equator.

---

[1] The instant of greatest eclipse occurs when the Moon's shadow axis passes closest to Earth's center.

[2] Terrestrial Time or TT is a time scale established by astronomers to calculate the location or planetary position of other objects as they relate to the center of Earth and its rotation.

[3] Coordinated Universal Time or UTC is the primary time standard that regulates world clocks and time. It is based on International Atomic Time (TAI) with leap seconds added at irregular intervals to compensate for the slowing of Earth's rotation. UTC is not adjusted for daylight saving time. It is effectively a successor to Greenwich Mean Time (GMT).

[4] Greatest duration is the time (and location) when the length of totality reaches its maximum value.

Overview Map: Path of Totality – Africa and Arabian Peninsula

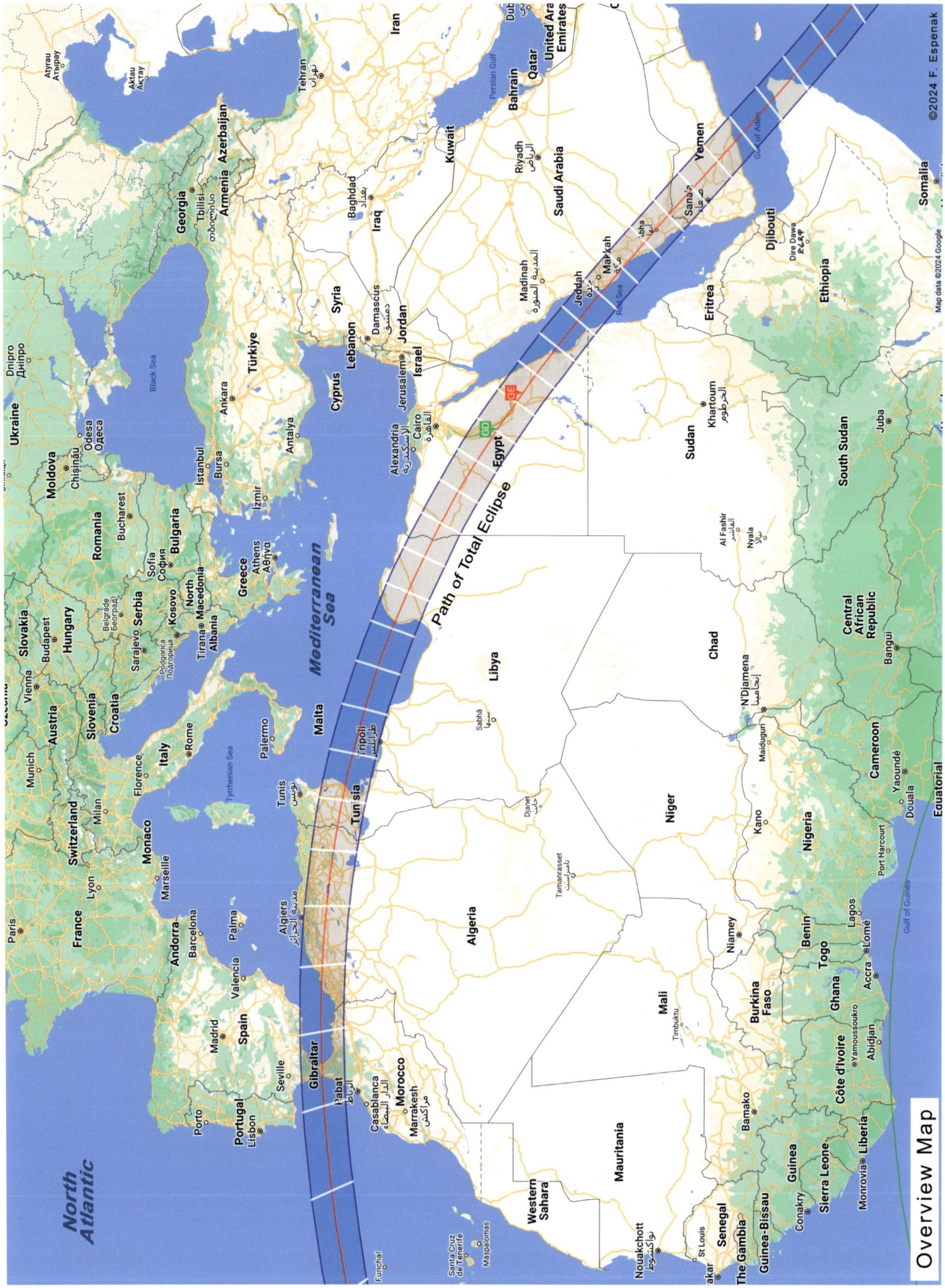

Figure 3 – Overview Map: Path of Totality – Africa and Arabian Peninsula

4 – Path of Totality

The path of the Moon's umbral shadow begins at 08:25 UTC in the central North Atlantic about 3500 km (2175 mi) east of Florida and 3300 km (2050 mi) from Africa. Along the sunrise terminator, the duration of totality is 3m 6s (i.e., 3 minutes 6 seconds) as seen from the center of the narrow, 206 km (128 mi) wide path. No major landfall occurs during the first 22 minutes of the central eclipse as the shadow sweeps northeast across the Atlantic.

Table 1 – Time Zones used by Countries within the 2027 Path of Totality

Time Zone Name	Abbrev.	UTC Offset	Countries in 2027 Path
Western European Summer Time	WEST	UTC + 1 hr	Portugal, Morocco
Central European Time	CET	UTC + 1 hr	Algeria, Tunisia
Central European Summer Time	CEST	UTC + 2 hrs	Spain, Gibraltar
Eastern European Time	EET	UTC + 2 hrs	Libya
Central Africa Time	CAT	UTC + 2 hrs	Sudan
Eastern European Summer Time	EEST	UTC + 3 hrs	Egypt
Arabia Standard Time	AST	UTC + 3 hrs	Saudi Arabia, Yemen
Eastern Africa Time	EAT	UTC + 3 hrs	Somalia

Figure 4 – The path of totality crosses the Strait of Gibraltar and includes parts of Spain, Morocco, Algeria, and Tunisia.

The umbra finally reaches terra firma at the Strait of Gibraltar (Figure 3 and Figure 4) at 08:47 UTC (09:47 am WEST). The Sun's altitude is 38° and the track width is 237 km (147 mi). Tangier, Morocco lies near the central line where the duration of totality is 4m 51s The southern tip of Spain also lies within the path where the nearby British Overseas Territory of Gibraltar has a duration of 4 min 28 sec.

Traveling with a velocity of 1.284 km/s (2873 mph), the umbral shadow quickly leaves Spain and runs along the Mediterranean coast of Morocco. At 08:53 UTC (09:53 am CET), the central line reaches the coast of Algeria as the entire path moves inland. Unfortunately, the city of Algiers lies about 15 km (9 mi) outside the northern edge of the path. Its 4.5 million citizens will experience a 0.997 magnitude partial eclipse.

At 09:07 UTC (10:07 am CET), the path leaves Algeria and enters Tunisia. The central duration has increased to 5m 37s and the Sun is 53° above the horizon. Four minutes later, the path reaches Tunisia's eastern coastline with the Mediterranean. The city of Sfax and its 1 million inhabitants will be treated to totality lasting 5m 44s.

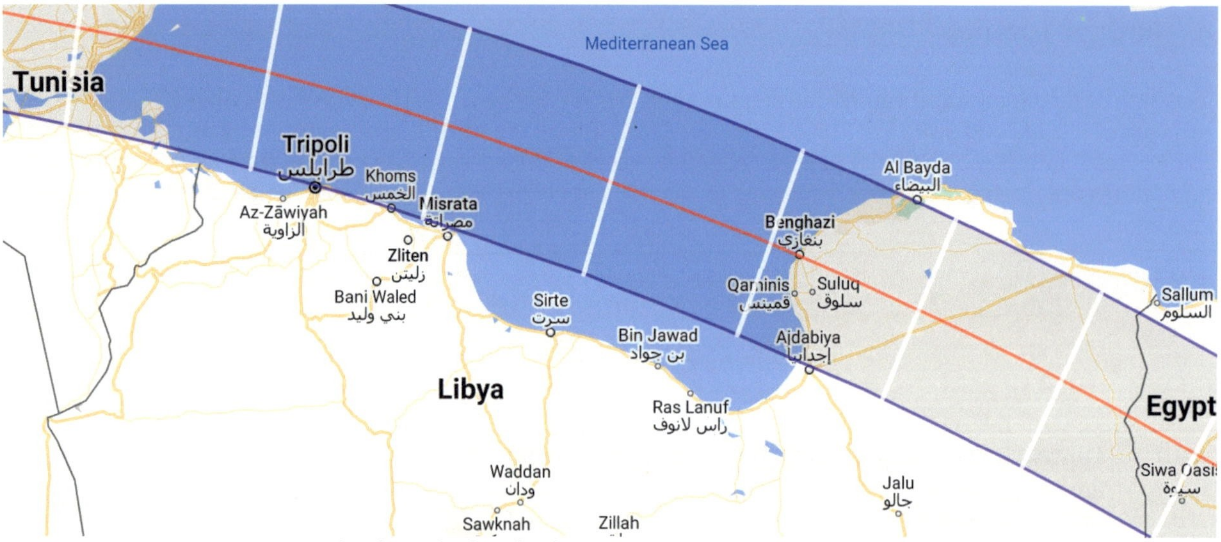

Figure 5 – The path of totality leaves Tunisia and crosses northern Libya.

Leaving Tunisia, the path returns to the Mediterranean where it clips the northern coast of Libya (Figure 5). The capital city of Tripoli lies just outside the southern limit of the path. The eclipse magnitude in Tripoli is a tantalizing 0.999 at 09:17 UTC (11:17 am EET). The path moves inland again in eastern Libya. Benghazi is on the central line for a duration of 6m 09s and a Sun altitude of 68° at 09:31 UTC (11:31 am EET).

The shadow track is on a southeastern course as it continues deeper into the Libyan dessert where it crosses the border of Egypt at 09:38 UTC (12:38 pm EEST). The duration has increased to 6m 15s.

Figure 6 – The path of totality runs diagonally across Egypt before reaching the Red Sea.

The 10 million people living in Cairo, Egypt's capital city, witness a partial eclipse of magnitude 0.946 at 09:56 UTC (12:56 pm EEST). About 190 km (118 mi) south lies the northern edge of the path of totality (Figure 6). A 500 km (310 mi) long stretch of the Nile River Valley meanders through the path. Almost as if it was planned, Luxor stands near the center. It is the site of the Ancient Egyptian city of Thebes and contains some of world's greatest archeological treasures including the Great Temple of Karnak and (across the Nile) the Valley of the Kings.

The duration of totality in Luxor is a generous 6m 21s centered at 10:05 UTC (1:05 pm EEST) with the Sun nearly overhead (altitude of 82°). Just 58 km (36 mi) to the southeast of Luxor lies the point of greatest eclipse at 10:06:41 UTC (1:06:41 pm EEST) with a duration of 6m 22.64s. But the point of greatest duration of totality occurs 6 minutes earlier (10:00:24 UTC) and 250 km (155 mi) to the northwest, just south of Asyut. Greatest duration is 6m 23.24s or 0.6s more than greatest eclipse.

The path continues southeast leaving Egypt and clipping a small corner of Sudan as it crosses the Red Sea and reaches the Arabian Peninsula (Figure 6). Southwestern Saudi Arabia is the next nation within the eclipse track. The port city of Jeddah with its 3.7 million inhabitants will witness totality at 10:25 UTC (1:25 pm AST) with a duration of 5m 57s. Seventy kilometers (45 mi) to the east lies Makkah (Mecca), the fountainhead and cradle of Islam. The duration of totality there is 5m 07s.

Figure 7 – The path crosses the Red Sea, Saudi Arabia, Yemen and Somalia.

At 10:40 UTC (1:40 pm AST), the track enters western Yemen. Sana'a is the largest Yemeni city within the path but it lies near the southern limit and experiences 2m 22s of totality. Continuing to Yemen's southern coast, the path begins its transit of the Gulf of Aden.

The Somali Peninsula (aka the Horn of Africa) is the path of totality's last landfall. The central line duration of totality here is 5m 30s and the Sun's altitude is 56°. Bosaso is the largest Somalian city (pop. 701,000) in the path with a duration of 4m 03s.

Leaving the Somalian coast at approximately 11:05 UTC (2:05 pm EAT), and with a velocity of 0.88 km/s (1971 mi/hr), the umbra races 5801 km (3605 mi) across the Indian Ocean before reaching the end of the track at 11:48 UTC and returning to space. Over a period of 3 hours 23 minutes, the umbra travels along a 15,227 km (9461 mi) track that covers an area equal to 0.71% of Earth's surface.

Photo 3 – Time sequence of the Total Solar Eclipse of August 21, 2017.
This sequence shows the partial phases, the diamond ring effect, and the solar corona during totality.
© 2017 F. Espenak, www.MrEclipse.com

Table 2 – The 2027 Path of Totality

UTC	Latitude	Longitude	Diameter Ratio	Sun Alt	Sun Azm	Path Width km	Major Axis km	Minor Axis km	Duration Totality
08:25.1	27°57.5'N	044°29.4'W	1.060	0.0°	69.8°	205.9	—	198.0	03m06.1s
08:30	33°03.6'N	026°01.0'W	1.066	17.7°	79.9°	221.6	711.6	215.6	03m50.8s
08:40	35°14.1'N	012°31.8'W	1.070	31.0°	89.4°	231.9	442.4	227.9	04m30.1s
08:50	35°44.4'N	003°26.3'W	1.073	40.4°	96.8°	238.4	363.4	235.6	04m58.9s
09:00	35°28.6'N	003°46.1'E	1.074	48.2°	103.6°	243.2	323.5	241.2	05m22.3s
09:10	34°43.5'N	009°48.9'E	1.076	55.2°	110.2°	247.0	299.2	245.5	05m41.4s
09:20	33°37.3'N	015°02.4'E	1.077	61.5°	117.1°	250.1	283.3	248.8	05m56.7s
09:30	32°14.7'N	019°38.7'E	1.078	67.3°	124.9°	252.6	272.5	251.3	06m00.4s
09:40	30°38.7'N	023°46.3'E	1.078	72.6°	135.0°	254.5	265.3	253.1	06m16.7s
09:50	28°51.3'N	027°31.3'E	1.079	77.4°	150.1°	256.0	260.7	254.3	06m21.6s
10:00	26°53.9'N	030°58.9'E	1.079	80.8°	176.4°	257.1	258.3	255.0	06m23.2s
10:10	24°47.1'N	034°13.4'E	1.079	81.5°	215.3°	258.0	258.0	255.1	06m21.9s
10:20	22°31.4'N	037°19.2'E	1.079	78.9°	246.6°	258.5	259.6	254.6	06m17.6s
10:30	20°06.8'N	040°20.3'E	1.079	74.4°	264.0°	258.8	263.3	253.5	06m10.5s
10:40	17°32.8'N	043°21.1'E	1.078	69.3°	273.9°	258.7	269.4	251.9	06m00.9s
10:50	14°48.4'N	046°26.9'E	1.077	63.6°	280.2°	258.2	278.8	249.6	05m48.7s
11:00	11°52.0'N	049°44.1'E	1.076	57.5°	284.5°	257.1	292.6	246.5	05m34.1s
11:10	08°40.7'N	053°22.3'E	1.075	50.8°	287.5v	254.9	313.2	242.6	05m16.8s
11:20	05°09.4'N	057°36.6'E	1.073	43.3°	289.6°	251.1	346.2	237.4	04m56.5s
11:30	01°07.8'N	062°57.3'E	1.071	34.6°	290.9°	244.7	406.4	230.5	04m32.3s
11:40	03°52.6'S	070°47.3'E	1.068	23.1°	291.2°	233.5	561.0	220.4	04m01.3s
11:48.3	12°29.1'S	090°25.6'E	1.060	0.0°	288.2°	205.3	—	197.5	03m05.6s

Table 2 lists the geographic coordinates of the central line for every ten minutes UTC. *Diameter Ratio* gives the Moon's apparent diameter divided by the Sun's. The Sun's altitude (*Alt*) and azimuth (*Azm*) are given along with the width of the path of totality (kilometers). The *major* and *minor* axes of the umbral shadow are also listed. Finally, the *duration of totality* is given (minutes and seconds).

5 – Local Circumstances for the Eclipse

Local circumstances for the ***Total Eclipse*** for over 70 cities are found in Table 3 (Gibraltar, Spain and Morocco), Table 4 (Algeria and Tunisia), Table 5 (Libya, Egypt and Somalia), and Table 6 (Saudi Arabia and Yemen). All times are given in local time (see Table 1 for Time Zones). The *Duration Totality* is the period between the beginning and end of totality. These times include the effects of the Moon's limb profile. The Sun's altitude is given at the instant of mid-totality.

Local circumstances for the ***Partial Eclipse*** for many cities are found in Table 7 (Europe), Table 8 (Africa), Table 9 (Middle East), and Table 10 (Asia). All times are given in local time and include *Daylight Saving Time* (aka *Summer Time* in Europe). The Sun's altitude is given at the instant of maximum eclipse. *Eclipse Magnitude* is given as the fraction of the Sun's *diameter* covered by the Moon at maximum eclipse.

The 2027 Total Solar Eclipse Circumstances Calculator is an interactive web page that can quickly calculate the local circumstances for the solar eclipse from any geographic location not included in Tables 3 through 10. The URL address for the calculator is:

https://www.eclipsewise.com/solar/SEcirc/2001-2100/SE2027Aug02Tcirc.html

Table 3 – Total Eclipse Circumstances for Gibraltar, Spain, and Morocco

Country/City	Partial Eclipse Begins	Total Eclipse Begins	Total Eclipse Ends	Partial Eclipse Ends	Duration Totality	Sun Altitude
Gibraltar						
Gibraltar	09:41:08 am	10:45:34 am	10:50:04 am	12:01:11 pm	04m28s	38°
Spain						
Algeciras	09:41:02 am	10:45:24 am	10:49:53 am	12:00:56 pm	04m27s	38°
Cadiz	09:40:46 am	10:45:26 am	10:48:12 am	11:59:36 am	02m52s	37°
El Puerto de S. Maria	09:40:51 am	10:45:45 am	10:48:07 am	11:59:43 am	02m26s	38°
Jerez de la Frontera	09:40:58 am	10:46:14 am	10:47:56 am	11:59:51 am	01m40s	38°
La Linea	09:41:10 am	10:45:39 am	10:50:04 am	12:01:14 pm	04m24s	38°
Malaga	09:42:07 am	10:48:08 am	10:50:02 am	12:02:39 pm	01m54s	39°
San Fernando	09:40:47 am	10:45:20 am	10:48:30 am	11:59:46 am	03m15s	38°
Morocco						
Tangier	08:40:41 am	09:44:47 am	09:49:37 am	11:00:33 am	04m50s	38°
Tetouan	08:40:48 am	09:45:10 am	09:49:58 am	11:01:10 am	04m50s	38°
Spanish Terr.						
Ceuta	08:41:01 am	09:45:20 am	09:50:09 am	11:01:15 am	04m49s	38°

Table 4 – Total Eclipse Circumstances for Algeria, and Tunisia

Country/City	Partial Eclipse Begins	Total Eclipse Begins	Total Eclipse Ends	Partial Eclipse Ends	Duration Totality	Sun Altitude
Algeria						
Ain Beida	08:52:33 am	10:03:35 am	10:07:49 am	11:24:23 am	04m17s	52°
Ain Temouchent	08:43:52 am	09:50:31 am	09:55:16 am	11:08:38 am	04m42s	43°
Barika	08:50:09 am	09:59:44 am	10:05:11 am	11:20:47 am	05m27s	50°
Batna	08:51:06 am	10:01:07 am	10:06:25 am	11:22:14 am	05m15s	51°
Biskra	08:50:24 am	10:00:34 am	10:05:38 am	11:21:49 am	05m03s	51°
Blida	08:47:59 am	09:57:11 am	10:00:01 am	11:15:25 am	02m53s	47°
Bordj Bou Arreridj	08:49:45 am	09:59:24 am	10:03:39 am	11:19:13 am	04m16s	49°
Boufarik	08:48:07 am	09:57:48 am	09:59:43 am	11:15:31 am	01m54s	47°
Bou Saada	08:48:50 am	09:58:01 am	10:03:14 am	11:18:37 am	05m16s	49°
Djelfa	08:47:42 am	09:57:27 am	10:00:53 am	11:17:04 am	03m24s	48°
Ech Cheliff	08:46:26 am	09:54:10 am	09:58:46 am	11:12:49 am	04m34s	45°
El Eulma	08:50:44 am	10:01:08 am	10:04:37 am	11:20:52 am	03m35s	50°
Ghilizane	08:45:28 am	09:52:37 am	09:57:48 am	11:11:28 am	05m11s	45°
Khemis	08:47:18 am	09:55:40 am	09:59:47 am	11:14:23 am	04m08s	46°
Khenchla	08:52:12 am	10:02:45 am	10:08:03 am	11:24:13 am	05m16s	52°
Lemdiyya	08:47:51 am	09:56:30 am	10:00:45 am	11:15:33 am	04m16s	47°
Maghniyya	08:43:06 am	09:50:25 am	09:53:17 am	11:07:27 am	02m47s	42°
Mascara	08:45:03 am	09:51:57 am	09:57:07 am	11:10:35 am	05m10s	44°
Mostaganem	08:45:11 am	09:52:08 am	09:57:15 am	11:10:44 am	05m07s	44°
M'Sila	08:49:22 am	09:58:34 am	10:03:43 am	11:18:56 am	05m07s	49°
Oran	08:44:25 am	09:50:58 am	09:56:04 am	11:09:16 am	05m08s	43°
Saida	08:44:48 am	09:52:58 am	09:56:13 am	11:11:06 am	03m12s	44°
Setif	08:50:30 am	10:00:45 am	10:04:16 am	11:20:25 am	03m37s	50°
Sidi Bel Abbes	08:44:40 am	09:51:52 am	09:56:33 am	11:10:25 am	04m38s	44°
Tebessa	08:53:20 am	10:04:28 am	10:09:32 am	11:26:05 am	05m03s	53°
Tiaret	08:46:08 am	09:53:50 am	09:58:57 am	11:13:08 am	05m11s	46°
Tlemcen	08:43:36 am	09:51:04 am	09:54:13 am	11:08:31 am	03m06s	43°
Tunisia						
El Kairouan	08:55:49 am	10:08:56 am	10:11:33 am	11:29:41 am	02m37s	55°
El Kasserine	08:54:03 am	10:05:24 am	10:10:52 am	11:27:35 am	05m26s	54°
Gabes	08:55:31 am	10:08:41 am	10:12:49 am	11:31:16 am	04m08s	56°
Gafsa	08:53:54 am	10:05:43 am	10:10:54 am	11:28:11 am	05m08s	54°
Houmt Essouk	08:56:29 am	10:09:30 am	10:14:36 am	11:32:40 am	05m02s	57°
Sfax	08:56:27 am	10:08:48 am	10:14:29 am	11:31:49 am	05m40s	56°
Zarzis	08:56:49 am	10:11:18 am	10:14:12 am	11:33:38 am	02m54s	57°

Table 5 – Total Eclipse Circumstances for Libya, Egypt and Somalia

Country/City	Partial Eclipse Begins	Total Eclipse Begins	Total Eclipse Ends	Partial Eclipse Ends	Duration Totality	Sun Altitude
Libya						
Benghazi	11:10:43 am	12:27:53 pm	12:34:01 pm	01:53:17 pm	06m09s	68°
Berkane	08:42:46 am	09:49:25 am	09:53:01 am	11:06:32 am	03m36s	42°
Ksar-El-Kebir	08:40:10 am	09:44:52 am	09:48:39 am	11:00:19 am	03m46s	38°
Larache	08:40:05 am	09:44:21 am	09:48:43 am	10:59:52 am	04m19s	38°
Nador	08:42:24 am	09:48:19 am	09:52:42 am	11:05:27 am	04m20s	41°
Oujda	08:43:07 am	09:51:30 am	09:52:31 am	11:07:38 am	01m07s	42°
Tangier	08:40:41 am	09:44:47 am	09:49:37 am	11:00:33 am	04m50s	38°
Tetouan	08:40:48 am	09:45:10 am	09:49:58 am	11:01:10 am	04m50s	38°
Egypt						
Abnub	11:35:23 am	12:56:45 pm	01:02:48 pm	02:21:19 pm	06m00s	80°
Abu Tij	11:35:59 am	12:57:21 pm	01:03:35 pm	02:22:00 pm	06m11s	81°
Akhmim	11:37:21 am	12:58:53 pm	01:05:14 pm	02:23:32 pm	06m21s	81°
Al-Manshah	11:37:36 am	12:59:10 pm	01:05:32 pm	02:23:49 pm	06m22s	81°
Al-Minya	11:33:45 am	12:55:39 pm	12:59:44 pm	02:19:08 pm	04m07s	80°
Armant	11:40:09 am	01:02:03 pm	01:08:16 pm	02:26:32 pm	06m15s	82°
Asyut	11:35:32 am	12:56:53 pm	01:03:03 pm	02:21:31 pm	06m06s	81°
Edfu	11:41:39 am	01:04:08 pm	01:09:39 pm	02:28:14 pm	05m32s	82°
Esna	11:40:34 am	01:02:49 pm	01:08:34 pm	02:27:08 pm	05m47s	82°
Jirja	11:37:56 am	12:59:33 pm	01:05:54 pm	02:24:12 pm	06m23s	81°
Kom Ombo	11:42:28 am	01:06:09 pm	01:09:44 pm	02:29:16 pm	03m41s	83°
Luxor	11:40:19 am	01:02:09 pm	01:08:28 pm	02:26:36 pm	06m21s	82°
Mallawi	11:34:16 am	12:55:43 pm	01:01:04 pm	02:19:57 pm	05m23s	80°
Manfalut	11:34:57 am	12:56:15 pm	01:02:24 pm	02:20:55 pm	06m06s	80°
Qina	11:39:54 am	01:01:35 pm	01:07:47 pm	02:25:50 pm	06m09s	81°
Samalut	11:33:28 am	12:55:51 pm	12:58:49 pm	02:18:42 pm	02m58s	79°
Sawhaj	11:37:18 am	12:58:50 pm	01:05:12 pm	02:23:30 pm	06m22s	81°
Tahta	11:36:39 am	12:58:06 pm	01:04:26 pm	02:22:47 pm	06m19s	81°
Tima	11:36:22 am	12:57:47 pm	01:04:03 pm	02:22:26 pm	06m15s	81°
Somalia						
Bosaso	12:39:43 pm	01:58:10 pm	02:02:17 pm	03:12:00 pm	04m05s	58°

Table 6 – Total Eclipse Circumstances for Saudi Arabia and Yemen

Country/City	Partial Eclipse Begins	Total Eclipse Begins	Total Eclipse Ends	Partial Eclipse Ends	Duration Totality	Sun Altitude
Saudi Arabia						
Abha	12:13:11 pm	01:34:17 pm	01:40:19 pm	02:53:49 pm	06m03s	71°
Taif	12:03:34 pm	01:26:03 pm	01:29:57 pm	02:45:45 pm	03m54s	75°
Jeddah	12:00:31 pm	01:22:23 pm	01:28:23 pm	02:43:49 pm	05m56s	76°
Khamis Mushayt	12:13:33 pm	01:34:34 pm	01:40:33 pm	02:53:56 pm	06m00s	70°
Makkah (Mecca)	12:01:58 pm	01:24:05 pm	01:29:02 pm	02:44:41 pm	05m01s	76°
Yemen						
Sana'a	12:21:54 pm	01:44:06 pm	01:46:39 pm	03:00:37 pm	02m36s	67°

Table 7 - Partial Eclipse Circumstances for Europe

Country	City	Partial Eclipse Begins	Maximum Eclipse	Partial Eclipse Ends	Eclipse Magnitude	Sun Altitude
Austria	Vienna	10:14:05 am	11:17:17 am	12:22:23 pm	0.562	53°
Belarus	Minsk	11:39:01 am	12:29:29 pm	01:20:11 pm	0.316	53°
Belgium	Brussels	10:05:43 am	11:04:03 am	12:05:05 pm	0.531	44°
Bulgaria	Sofia	11:18:23 am	12:29:01 pm	01:40:45 pm	0.676	62°
Croatia	Zagreb	10:10:32 am	11:17:08 am	12:25:59 pm	0.640	54°
Cyprus	Nicosia	11:35:21 am	12:52:58 pm	02:08:37 pm	0.770	73°
Czech Rep.	Prague	10:14:31 am	11:14:54 am	12:17:09 pm	0.516	50°
Denmark	Copenhagen	10:22:59 am	11:13:42 am	12:05:37 pm	0.356	46°
UK	Birmingham	09:04:29 am	09:59:17 am	10:56:44 am	0.488	39°
	Glasgow	09:10:33 am	09:59:43 am	10:50:53 am	0.386	36°
	Leeds	09:07:24 am	10:00:19 am	10:55:34 am	0.447	38°
	Liverpool	09:05:44 am	09:59:00 am	10:54:44 am	0.460	38°
	London	09:03:35 am	10:00:08 am	10:59:29 am	0.518	40°
	Manchester	09:06:18 am	09:59:34 am	10:55:17 am	0.458	38°
	Newcastle	09:09:56 am	10:01:03 am	10:54:16 am	0.412	38°
Finland	Helsinki	11:49:02 am	12:25:34 pm	01:02:16 pm	0.168	46°
France	Lille	10:04:20 am	11:02:43 am	12:03:56 pm	0.540	43°
	Lyon	09:58:14 am	11:02:55 am	12:11:23 pm	0.693	46°
	Marseille	09:55:48 am	11:03:03 am	12:14:35 pm	0.771	47°
	Paris	10:00:44 am	11:01:14 am	12:05:01 pm	0.598	43°
Germany	Berlin	10:17:27 am	11:13:54 am	12:11:58 pm	0.448	48°
	Cologne	10:08:11 am	11:06:47 am	12:07:51 pm	0.521	45°
	Essen	10:09:10 am	11:06:59 am	12:07:10 pm	0.504	45°
	Frankfurt	10:08:22 am	11:08:21 am	12:10:49 pm	0.541	47°
	Hamburg	10:15:57 am	11:10:37 am	12:07:01 pm	0.430	46°
	Munich	10:08:17 am	11:11:22 am	12:16:59 pm	0.591	50°
	Stuttgart	10:06:42 am	11:08:38 am	12:13:16 pm	0.582	48°
Greece	Athens	11:16:48 am	12:32:56 pm	01:50:03 pm	0.816	66°
Hungary	Budapest	10:16:47 am	11:20:59 am	12:26:44 pm	0.566	55°
Iceland	Reykjavik	08:30:47 am	08:59:11 am	09:28:07 am	0.131	25°
Ireland	Dublin	09:03:51 am	09:56:26 am	10:51:36 am	0.463	36°
Italy	Milan	10:02:09 am	11:08:11 am	12:17:35 pm	0.687	50°
	Naples	10:03:43 am	11:15:43 am	12:31:00 pm	0.809	56°
	Rome	10:02:13 am	11:12:47 am	12:26:49 pm	0.786	54°
Latvia	Riga	12:39:54 pm	01:25:05 pm	02:10:37 pm	0.258	49°
Lithuania	Vilnius	12:37:01 pm	01:26:46 pm	02:16:55 pm	0.312	51°
Macedonia	Skopje	10:14:53 am	11:26:28 am	12:39:36 pm	0.716	61°
Malta	Valletta	10:01:47 am	11:17:52 am	12:37:46 pm	0.968	60°
Netherlands	Amsterdam	10:09:02 am	11:05:14 am	12:03:45 pm	0.482	43°
Norway	Oslo	10:31:39 am	11:13:26 am	11:55:59 am	0.240	42°
Poland	Katowice	10:20:33 am	11:20:21 am	12:21:28 pm	0.483	53°
	Warsaw	10:26:28 am	11:22:26 am	12:19:17 pm	0.411	52°
Portugal	Lisbon	08:40:48 am	09:44:53 am	10:55:07 am	0.929	35°
Romania	Bucharest	11:24:23 am	12:32:11 pm	01:40:29 pm	0.597	61°
Russia	Gorky	12:14:54 pm	12:45:33 pm	01:15:57 pm	0.108	51°
	Moscow	12:00:01 pm	12:39:26 pm	01:18:36 pm	0.182	52°
	Niz.Novgorod	12:14:54 pm	12:45:33 pm	01:15:57 pm	0.108	51°
	St. Petersburg	11:56:29 am	12:30:21 pm	01:04:15 pm	0.140	47°
Serbia & Mont.	Belgrade	10:15:50 am	11:23:46 am	12:33:16 pm	0.636	58°
Spain	Barcelona	09:50:55 am	10:58:34 am	12:11:18 pm	0.842	45°
	Bilbao	09:49:00 am	10:53:17 am	12:02:32 pm	0.785	40°
	Madrid	09:45:29 am	10:51:17 am	12:02:48 pm	0.880	40°
	Sevilla	09:41:31 am	10:47:33 am	12:00:01 pm	0.979	38°
	Valencia	09:47:06 am	10:54:52 am	12:08:24 pm	0.910	43°
	Zaragoza	09:48:41 am	10:54:56 am	12:06:27 pm	0.838	42°
Sweden	Stockholm	10:37:48 am	11:19:25 am	12:01:32 pm	0.227	45°
Switzerland	Zurich	10:04:00 am	11:07:37 am	12:14:18 pm	0.628	48°
Ukraine	Kiev	12:38:30 pm	01:34:35 pm	02:30:38 pm	0.384	57°

Table 8 - Partial Eclipse Circumstances for Africa

Country	City	Partial Eclipse Begins	Maximum Eclipse	Partial Eclipse Ends	Eclipse Magnitude	Sun Altitude
Algeria	Algiers	08:48:21 am	09:58:59 am	11:15:39 am	0.995	47°
	Annaba	08:53:24 am	10:06:10 am	11:24:13 am	0.975	52°
	Constantine	08:51:52 am	10:04:24 am	11:22:29 am	0.998	51°
Benin	Cotonou	09:41:43 am	10:11:16 am	10:41:56 am	0.064	48°
Burkina Faso	Ouagadougou	08:06:16 am	08:56:23 am	09:51:10 am	0.247	43°
Burundi	Bujumbura	11:57:08 am	12:49:37 pm	01:38:59 pm	0.194	67°
Cameroon	Douala	10:09:27 am	10:37:01 am	11:04:58 am	0.046	60°
	Yaounde	10:11:35 am	10:43:27 am	11:15:42 am	0.061	63°
C.A.R.	Bangui	10:12:51 am	11:06:04 am	11:58:36 am	0.179	73°
Chad	Ndjamena	09:34:33 am	10:42:11 am	11:51:35 am	0.351	69°
D.R.C.	Kisangani	11:42:22 am	12:34:20 pm	01:23:55 pm	0.177	72°
Djibouti	Djibouti	12:27:16 pm	01:50:29 pm	03:05:12 pm	0.887	66°
Egypt	Alexandria	11:29:41 am	12:51:41 pm	02:12:01 pm	0.935	76°
	Al-Mahallah Al-Kubra	11:32:26 am	12:54:28 pm	02:14:22 pm	0.921	77°
	Aswan	11:42:50 am	01:08:26 pm	02:29:51 pm	0.995	83°
	Bahtim	11:33:14 am	12:55:51 pm	02:16:09 pm	0.943	78°
	Cairo	11:33:13 am	12:55:54 pm	02:16:16 pm	0.946	78°
	Giza	11:33:10 am	12:55:53 pm	02:16:16 pm	0.948	78°
	Ismailia	11:35:00 am	12:57:08 pm	02:16:42 pm	0.914	77°
	Port Said	11:34:39 am	12:56:14 pm	02:15:23 pm	0.895	76°
	Shubra al-Khaymah	11:33:11 am	12:55:50 pm	02:16:10 pm	0.945	78°
	Suez	11:36:02 am	12:58:34 pm	02:18:20 pm	0.927	78°
	Tanta	11:32:12 am	12:54:24 pm	02:14:28 pm	0.929	77°
Eritrea	Asmera	12:10:50 pm	01:36:06 pm	02:53:51 pm	0.887	74°
Ethiopia	Addis Ababa	12:24:52 pm	01:47:20 pm	03:01:52 pm	0.723	70°
Gambia	Banjul	07:44:36 am	08:31:35 am	09:23:26 am	0.327	23°
Ghana	Accra	08:44:42 am	09:04:47 am	09:25:16 am	0.030	44°
Guinea	Conakry	07:57:36 am	08:35:57 am	09:17:28 am	0.181	25°
Guinea-Bissau	Bissau	07:49:00 am	08:33:01 am	09:21:18 am	0.267	23°
Ivory Coast	Abidjan	08:40:21 am	08:55:33 am	09:10:53 am	0.019	38°
Kenya	Mombasa	01:03:51 pm	02:10:30 pm	03:11:00 pm	0.412	57°
	Nairobi	12:52:27 pm	02:01:30 pm	03:04:21 pm	0.417	62°
Liberia	Monrovia	08:16:40 am	08:41:41 am	09:07:47 am	0.064	29°
Libya	Tripoli	10:59:43 am	12:16:52 pm	01:38:27 pm	0.999	60°
Madagascar	Antananarivo	01:53:11 pm	02:35:55 pm	03:15:58 pm	0.193	36°
Malawi	Blantyre	01:05:29 pm	01:17:05 pm	01:28:38 pm	0.011	50°
Mali	Bamako	07:56:06 am	08:43:33 am	09:35:45 am	0.262	33°
Mauritania	Nouakchott	07:38:19 am	08:32:06 am	09:32:09 am	0.476	24°
Morocco	Casablanca	08:38:30 am	09:44:24 am	10:57:25 am	0.969	36°
	Fes	08:40:25 am	09:47:40 am	11:01:58 am	0.981	39°
	Marrakech	08:37:39 am	09:43:17 am	10:56:19 am	0.902	36°
	Meknes	08:39:53 am	09:46:47 am	11:00:48 am	0.975	38°
	Rabat	08:39:07 am	09:45:22 am	10:58:41 am	0.981	37°
Mozambique	Nacala	12:41:49 pm	01:24:31 pm	02:04:36 pm	0.165	46°
	Nampula	12:45:58 pm	01:23:01 pm	01:58:10 pm	0.120	47°
Niger	Niamey	09:08:27 am	10:04:07 am	11:04:52 am	0.292	48°
Nigeria	Ibadan	09:37:01 am	10:13:20 am	10:51:27 am	0.098	50°
	Lagos	09:42:28 am	10:13:56 am	10:46:39 am	0.071	50°
Rwanda	Kigali	11:50:35 am	12:48:51 pm	01:43:17 pm	0.246	68°
Senegal	Dakar	07:41:52 am	08:30:40 am	09:24:44 am	0.371	22°
Sierra Leone	Freetown	08:01:48 am	08:36:57 am	09:14:41 am	0.145	26°
Somalia	Mogadishu	12:52:59 pm	02:10:00 pm	03:18:40 pm	0.697	56°
Sudan	Al-Khartum	10:57:16 am	12:22:38 pm	01:42:15 pm	0.759	83°
	Port Sudan	10:59:13 am	12:25:03 pm	01:44:26 pm	0.962	79°
Tanzania	Dar es Salaam	01:12:50 pm	02:13:40 pm	03:09:19 pm	0.333	54°
Togo	Tunis	08:56:15 am	10:10:07 am	11:28:45 am	0.967	55°
Uganda	Kampala	12:42:52 pm	01:50:12 pm	02:52:17 pm	0.357	68°

Table 9 - Partial Eclipse Circumstances for the Middle East

Country	City	Partial Eclipse Begins	Maximum Eclipse	Partial Eclipse Ends	Eclipse Magnitude	Sun Altitude
Iran	Isfahan	01:49:06 pm	02:55:55 pm	03:58:19 pm	0.518	62°
	Mashhad	02:08:13 pm	02:59:52 pm	03:48:40 pm	0.305	54°
	Tehran	01:47:23 pm	02:50:39 pm	03:50:15 pm	0.456	62°
Iraq	Baghdad	12:01:01 pm	01:14:08 pm	02:23:05 pm	0.633	69°
Israel	Jerusalem	11:40:41 am	01:00:59 pm	02:18:10 pm	0.833	76°
	Tel Aviv	11:39:30 am	12:59:42 pm	02:16:58 pm	0.832	75°
Jordan	'Amman	11:42:08 am	01:01:59 pm	02:18:36 pm	0.816	75°
Kuwait	Kuwait City	12:12:14 pm	01:26:20 pm	02:35:04 pm	0.662	66°
Lebanon	Beirut	11:40:18 am	12:58:27 pm	02:13:53 pm	0.771	74°
Qatar	Doha	12:24:35 pm	01:38:40 pm	02:46:27 pm	0.686	62°
Saudi Arabia	Riyadh	12:13:44 pm	01:32:39 pm	02:45:01 pm	0.797	67°
Syria	Damascus	11:42:12 am	01:00:24 pm	02:15:37 pm	0.768	74°
U.A.E.	Abu Dhabi	01:32:14 pm	02:44:00 pm	03:49:28 pm	0.648	58°
Yemen	'Adan	12:29:00 pm	01:51:44 pm	03:05:55 pm	0.961	64°

Table 10 - Partial Eclipse Circumstances for Asia

Country	City	Partial Eclipse Begins	Maximum Eclipse	Partial Eclipse Ends	Eclipse Magnitude	Sun Altitude
Afghanistan	Kabul	02:33:56 pm	03:12:50 pm	03:49:55 pm	0.183	44°
Armenia	Jerevan	01:00:01 pm	02:04:03 pm	03:05:32 pm	0.468	65°
Azerbaijan	Baku	01:12:58 pm	02:11:09 pm	03:06:45 pm	0.380	61°
Bahrain	Al-Manamah	12:21:21 pm	01:35:39 pm	02:43:50 pm	0.683	63°
Bangladesh	Dacca	04:49:05 pm	05:12:49 pm	05:35:51 pm	0.090	19°
Comoros	Moroni	01:30:07 pm	02:25:06 pm	03:15:39 pm	0.296	46°
Georgia	Tbilisi	01:01:03 pm	02:02:24 pm	03:01:31 pm	0.427	63°
India	Ahmadabad	03:44:49 pm	04:35:10 pm	05:21:52 pm	0.349	36°
	Bangalore	03:58:57 pm	04:53:22 pm	05:43:13 pm	0.493	26°
	Bombay	03:47:26 pm	04:41:50 pm	05:31:49 pm	0.431	34°
	Calcutta	04:14:30 pm	04:43:58 pm	05:12:16 pm	0.138	20°
	Delhi	03:53:45 pm	04:29:06 pm	05:02:47 pm	0.168	34°
	Madras	04:02:47 pm	04:54:23 pm	05:41:54 pm	0.451	23°
Kazakhstan	Cimkent	04:10:24 pm	04:29:56 pm	04:49:11 pm	0.045	44°
Laos	Vientiane	05:59:25 pm	06:21:10 pm	06:42:24 pm	0.091	4°
Maldives	Male	03:29:47 pm	04:32:41 pm	05:29:22 pm	0.774	25°
Mauritius	Port Louis	02:53:41 pm	03:44:36 pm	04:31:38 pm	0.346	26°
Myanmar	Mandalay	05:26:53 pm	05:46:10 pm	06:05:02 pm	0.061	12°
	Yangon	05:21:55 pm	05:52:52 pm	06:22:31 pm	0.176	9°
Nepal	Kathmandu	04:12:28 pm	04:35:02 pm	04:57:02 pm	0.075	26°
Pakistan	Karachi	03:01:41 pm	03:57:29 pm	04:48:58 pm	0.404	43°
	Lahore	03:17:01 pm	03:52:07 pm	04:25:39 pm	0.157	38°
Seychelles	Victoria	02:20:11 pm	03:30:39 pm	04:33:29 pm	0.738	39°
Sri Lanka	Colombo	04:05:23 pm	05:01:56 pm	05:53:30 pm	0.608	20°
Tajikistan	Dusanbe	04:03:53 pm	04:35:29 pm	05:06:02 pm	0.118	45°
Thailand	Bangkok	05:54:03 pm	06:26:12 pm	-	0.206	4°
Turkey	Ankara	11:34:28 am	12:46:32 pm	01:57:28 pm	0.645	68°
	Bursa	11:27:09 am	12:40:06 pm	01:52:46 pm	0.687	67°
	Istanbul	11:27:18 am	12:39:11 pm	01:50:53 pm	0.664	66°
	Izmir	11:23:03 am	12:38:30 pm	01:54:01 pm	0.762	68°
Turkmenistan	Aschabad	02:34:51 pm	03:25:41 pm	04:13:57 pm	0.293	55°
Uzbekistan	Taskent	03:07:49 pm	03:31:17 pm	03:54:16 pm	0.065	45°

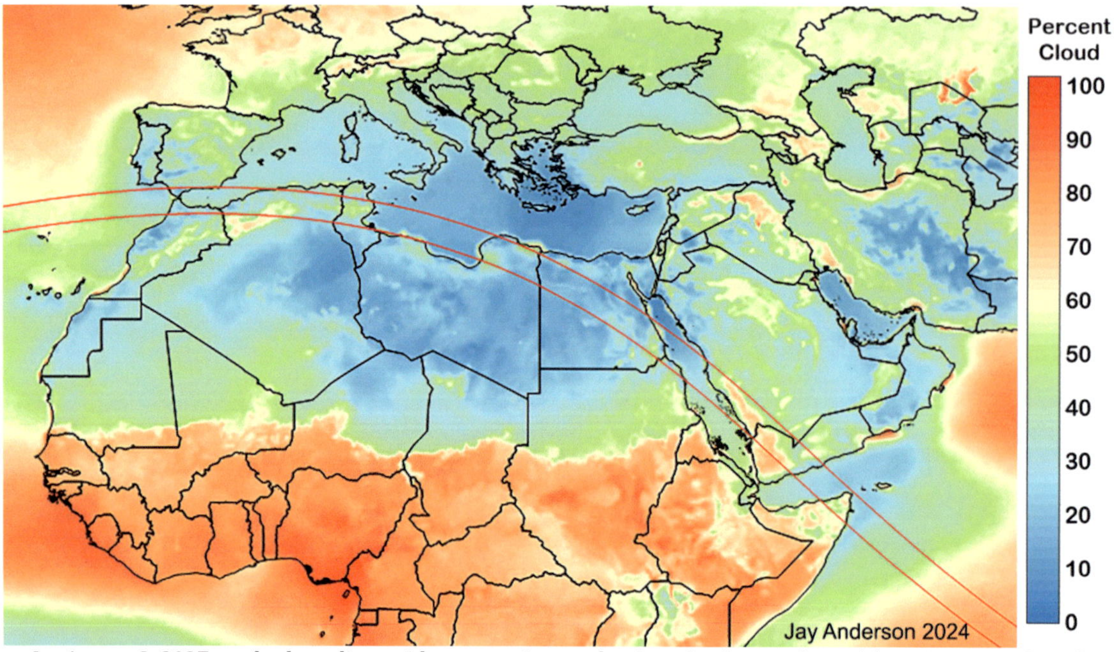

Figure 8 – August 2, 2027 total solar eclipse with average August daytime cloud cover derived from 20 years of satellite data, 2000-2019. Data: CM-SAF. (map by Jay Anderson)

6 – Weather Prospects for the 2027 Total Solar Eclipse

The track of the 2027 total eclipse crosses southernmost Spain, North Africa and the Arabian Peninsula. These regions all enjoy an arid climate and offer excellent prospects for cloud free sky during the eclipse.

Table 11 lists twenty cities in the path of totality and include the duration of totality, the average high temperature, and the average cloud cover during August. Some locations in Algeria, Tunisia, and Egypt have average cloud amounts of under 10% making them particularly desirable.

Table 11 – Average High Temperature and Cloud Cover in August

Country	City	Duration Totality	Ave. High Temp.[1]	Ave. Cloud Cover[2]
Spain	Cadiz	2m 52s	29°C (84°F)	13%
Spain	Malaga	1m 54s	29°C (84°F)	10%
Gibraltar	Gibraltar	4m 28s	27°C (81°F)	16%
Morocco	Tangier	4m 50s	30°C (86°F)	42%
Morocco	Tetouan	4m 50s	30°C (86°F)	25%
Algeria	Oran	5m 08s	31°C (88°F)	35%
Algeria	Blida	2m 53s	33°C (91°F)	8%
Algeria	Ain Batna	5m 17s	34°C (93°F)	7%
Algeria	Tebessa	5m 13s	33°C (92°F)	7%
Tunisia	Sfax	5m 40s	33°C (89°F)	2%
Libya	Benghazi	6m 09s	31°C (87°F)	23%
Egypt	Asyut	6m 07s	36°C (97°F)	0%
Egypt	Qina	6m 11s	41°C (105°F)	0%
Egypt	Luxor	6m 21s	40°C (104°F)	0%
Saudi Arabia	Jeddah	6m 00s	38°C (100°F)	27%
Saudi Arabia	Mecca	5m 08s	41°C (106°F)	4%
Saudi Arabia	Abha	6m 04s	28°C (82°F)	13%
Saudi Arabia	Abu Arish	4m 31s	39°C (102°F)	33%
Yemen	Sana'a	2m 20s	28°C (82°F)	23%
Somalia	Bosaso	4m 00s	37°C (98°F)	76%

[1]Average High Temperature in August [source: www.weatherspark.com]

[2]Average Cloud Cover in August (since 2000) [source: www.timeanddate.com]

7 – Saros 136

The total eclipse of 2027 is part of a series of eclipses with the longest totalities of the 20th and 21st centuries. These remarkable eclipses all have durations exceeding six minutes and belong to a group called Saros 136.

The Saros is a period in which eclipses repeat with very similar characteristics. It has a duration of approximately 6,585.3 days (~18 years 11 days 8 hours). When two eclipses are separated by a period of one Saros, they share a very similar geometry. The eclipses occur at the same node with the Moon at nearly the same distance from Earth and at the same time of year. Thus, the Saros is useful for organizing eclipses into families or series. A Saros series doesn't last forever because the characteristics of the eclipse geometry slowly change with each cycle. A series typically lasts 12 or 13 centuries and contains 70 or more eclipses.

The total eclipse of 2027 is the 38th member of Saros series 136. The Saros series were first numbered by van den Bergh in 1955. All eclipses in an even numbered Saros series occur at the Moon's descending node and the Moon moves northward with each succeeding member in the family (i.e., gamma[5] increases). Saros 136 is a vigorous series in the prime of its life. The series began with a small partial eclipse visible off the coast of Antarctica on Jun 14, 1360. After seven more partial eclipses, each of increasing magnitude, the first umbral eclipse occurred on Sep 8, 1504. This event was a central annular eclipse of short duration that was visible from Antarctica and the South Pacific Ocean.

The next five members of Saros 136 were all annular eclipses; the duration of annularity steadily decreased with each eclipse as the Moon passed progressively closer to Earth. The character of the series changed to hybrid (also called annular-total) with the eclipse of Nov 22, 1612. The nature of such an eclipse changes from annular to total or vice versa along different portions of the ground track. This dual nature arises due to the curvature of Earth's surface, which brings the middle part of the path into the umbra (total eclipse) while other, more distant segments remain within the antumbral shadow (annular eclipse). Hybrid eclipses are rather rare and account for only 4.8% of the 11,898 solar eclipses occurring during the 5000-year period from 2000 BCE to 3000 CE.

Detailed calculations reveal that the 1612 hybrid eclipse did not actually have a total phase. At greatest eclipse the Sun and Moon subtended virtually the same angular diameter, so the Moon's irregular profile produced a beaded annularity not unlike the solar eclipse of Oct 3, 1986. Beginning with the eclipse of Dec 4, 1630, the following five events were true hybrids with increasing durations of totality. The last of the hybrids in this group occurred on Jan 17, 1703. It is interesting to note that the entire path was total except for the final ~1500 km (930 mi) which was annular. In comparison, most hybrid eclipses are annular at both ends of their paths.

The first purely total eclipse of the series occurred on Jan 27, 1721 and had a maximum duration of 1m 07s. The central track extended from the South Pacific to the South Atlantic, crossing southernmost South America in the process. Succeeding eclipses in the series had increasingly longer durations as the paths shifted north. The total eclipse of Mar 24, 1811 had a maximum duration of 3m 27s, while the eclipse of May 18, 1901 was nearly twice as long at 6m 29s.

Throughout the 20th century, Saros 136 continued to produce exceptionally long total eclipses with tracks passing across the equator. The eclipse of May 29, 1919 is significant because it was the first eclipse used to measure the gravitational deflection of starlight by the Sun as predicted by Einstein's general theory of relativity.

On Jun 8, 1937, the central line duration exceeded 7 minutes for the first time for any total eclipse since the year 1098. Unfortunately, most of the path was over the Pacific Ocean. The following eclipse (Jun 20, 1955) was 4 seconds longer with a maximum duration of 7m 08s. Its track crossed Sri Lanka, Southeast Asia, and the Philippines. This event marked the maximum duration of all total eclipses belonging to Saros 136.

Subsequent members in the series are shorter as the Moon recedes farther from perigee. Nevertheless, Saros 136 will continue to produce long eclipses for some time to come. The first successor to the 1955 eclipse occurred on Jun 30, 1973 and lasted 7m 04s from North Africa. It was followed by the total eclipse of Jul 11,

---

[5] *Gamma is the distance of the Moon's shadow axis from Earth's center in units of equatorial Earth radii. It is defined at the instant of greatest eclipse when its absolute value is at a minimum.*

1991 with a duration of 6m 53s and a path through Hawaii, Mexico, Central and South America. The most recent eclipse of the series occurred on Jul 22, 2009 with a duration of 6m 39s from China and the Pacific Ocean.

In comparison, the Aug 2, 2027 and lasts a maximum of 6m 23s. The next total eclipse of Saros 136 occurs on Aug 12, 2045 (duration 6m 06s). It is noteworthy because its track passes over the United States from northern California to central Florida where the event will be witnessed by millions of people.

The paths of successive eclipses have steadily shifted north, but this trend reverses starting with the total eclipse of Aug 24, 2063. The retrograde effect is due to the inclination of Earth's axis as the date of each eclipse shifts ~11 days forward through the fall–winter season. The temporary downward shift is the result of the tipping of Earth's Northern Hemisphere away from the Sun, which is greater than the northern shift of the Moon with respect to the node. Nearly two centuries later, the path resumes its northern shift with the eclipse of Dec 12, 2243. By this time, the maximum duration of totality will drop to 3m 30s.

The series continues to produce total eclipses for two more centuries as the duration of totality gradually dwindles. The last total eclipse occurs on May 13, 2496 and has a maximum duration of 1m 02s. The final seven eclipses in Saros 136 are all partial events visible from the Northern Hemisphere. The series terminates with a partial eclipse visible primarily from eastern Siberia on Jul 30, 2622.

In summary, Saros series 136 includes 71 eclipses. It begins with 8 partial eclipses, followed by 6 annular, 6 hybrid, and 44 total eclipses. It ends with a string of 7 partial eclipses. From start to finish, the series spans a period of 1262 years. The EclipseWise.com catalog for Saros 136 has more details and maps for this series:

www.eclipsewise.com/solar/SEsaros/SEsaros136.html

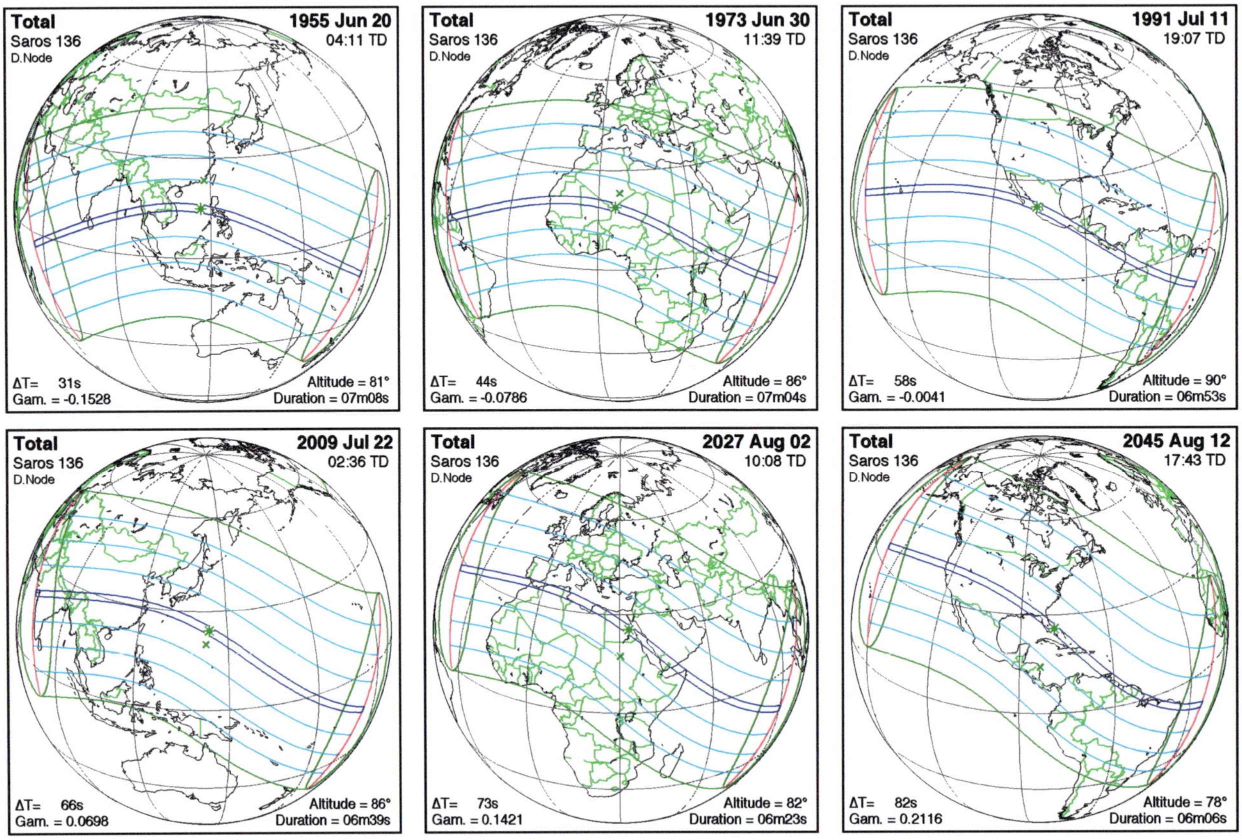

Figure 9 – Six total solar eclipses of Saros 136 for the years 1955 (7m 08s), 1973 (7m 04s), 1991 (6m 53s), 2009 (6m 39s), 2027 (6m 23s), and 2045 (6m 06s). Note how the eclipse path returns to approximately the same geographic region after 54 years, 33 days (3 cycles). This period is called an Exeligmos.

8 – Detailed Maps of the Path of Totality

A series of 20 full-page maps (pages 25-44) cover the entire 2027 path of totality through Spain, North Africa, and the Arabian Peninsula. The *Table of Contents* (page iii) can be used to quickly navigate to the map of interest since it lists the region featured on each map. North is up in each Map.

The map scale is approximately 1:2,500,000, which corresponds to 1 inch ≈ 40 miles (1 cm ≈ 25 km).[6] This large scale shows both major and minor roads, towns and cities, rivers, lakes, parks, national forests, wilderness areas and mountain ranges. A 50 mile (50 km) reference scale appears at the bottom of each map.

Five additional maps focus in on regions special interest and use a higher resolution of about 1:1,250,000, corresponding to 1 inch ≈ 20 miles (1 cm ≈ 13 km).

The path of totality on each map is depicted as a lightly shaded region with the northern and southern limits clearly labeled. The total eclipse can be seen only inside this path (a partial eclipse is visible outside the path). The closer one gets to the central line, the longer the total eclipse lasts. Gray lines inside the path mark the duration of the total eclipse in 30-second steps. This makes it easy to estimate the duration of totality from any location in the eclipse path.

The local time of mid-eclipse is marked by a series of white lines crossing the eclipse path every 5 minutes. Abbreviations for local times are: WEST, CET, CEST, EET, CAT,EEST, AST, and EAT (see Table 1). Eclipse circumstances on the central line are labeled with the local time of mid-eclipse, duration of totality (minutes and seconds) and altitude of the Sun.

All maps were produced using Google Maps as the underlying map with overlying eclipse graphics generated using JavaScript code. A web page is available to the user for examining any part of the 2027 eclipse path at a range of zoom magnifications. An added benefit of the web page is that it automatically calculates the local circumstances for any point chosen by the user, with just the click of a mouse. For more information and to access the interactive 2027 eclipse path plotted on Google Maps, visit:

https://www.eclipsewise.com/solar/SEgmapx/2001-2100/SE2027Aug02Tgmapx.html#map

9 – EclipseWise.com Web Site

EclipseWise.com has individual web pages, maps and diagrams for every solar and lunar eclipse covering the 5000-year period from 2000 BCE to 3000 CE. This includes 11,898 solar eclipses and 12,064 lunar eclipses. Much of the design, layout and graphics were inspired by the recent publications *Thousand Year Canon of Solar Eclipses 1501 to 2500* and the *Thousand Year Canon of Lunar Eclipses 1501 to 2500*. For more information on this book, see: *https://www.astropixels.com/pubs*

The graphical user interface used by *EclipseWise.com* offers an intuitive way of accessing eclipse predictions. For example, the home page presents a concise preview of all upcoming solar and lunar eclipses over several years. Each small eclipse diagram gives a quick preview of an eclipse and links to a dedicated page for that eclipse.

The main or top pages of EclipseWise.com are:

Home Page (both solar and lunar eclipses): ***https://www. eclipsewise.com/eclipse.html***
Solar Eclipses Page: ***https://www. eclipsewise.com/solar/solar.html***
Lunar Eclipses Page: ***https://www. eclipsewise.com/lunar/lunar.html***

[6] Because of the Mercator map projection, the actual scale on a given map can vary by up to 10% of this value.

10 – EclipseWise.com and the 2027 Eclipse

EclipseWise.com has a series of pages and resources devoted to the 2027 eclipse. The main page is located at:

https://www.eclipsewise.com/solar/SEprime/2001-2100/SE2027Aug02Tprime.html

It provides links to detailed eclipse path maps, tables of eclipse path coordinates, Besselian elements, and more. The link to an interactive Google Map with the eclipse path plotted on it allows the user to zoom into a part of the path. Click on any point on the map to display the eclipse circumstances and duration of totality at that location.

Other features include information on eye safety, eclipse photography, the sky during totality, and additional data tables about the eclipse path. This web site will continue to add features as the eclipse approaches.

The *2027 Total Solar Eclipse Circumstances Calculator* is an interactive web page that can quickly calculate the local circumstances for the eclipse from any geographic location. The *Calculator* is located at:

https://www.eclipsewise.com/solar/SEcirc/2001-2100/SE2027Aug02Tcirc.html

11 – Eclipse Predictions

The algorithms and software for the eclipse predictions were developed primarily from the *Explanatory Supplement to the Astronomical Ephemeris* (Her Majesty's Nautical Almanac Office, 1974) with additional algorithms from *Elements of Solar Eclipses: 1951–2200* (Meeus, 1989). The solar and lunar ephemerides were generated from the JPL DE430. All eclipse calculations were made using a value for the Moon's radius of k=0.2722810 for the path of totality. Eclipse contact times for locations and cities (Tables 3 through 6) have been corrected for the effects of the Moon's limb profile. A value for ΔT of 69.2 seconds was used to convert the predictions from Terrestrial Time (TT) to Coordinated Universal Time (UTC).

Photo 4 – Viewing the Total Solar Eclipse of April 08, 2024.
Mazatlán, Mexico, © 2024 F. Espenak, www.MrEclipse.com

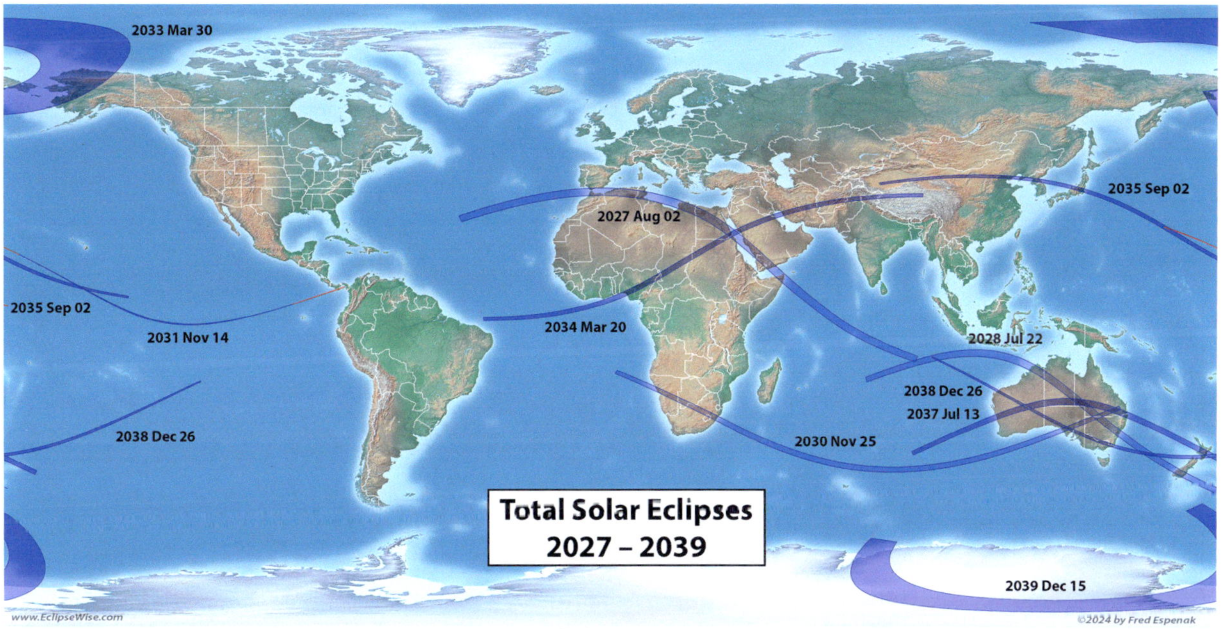

Figure 10 – The path of every total solar eclipse from 2027 through 2039 is plotted on a world map.

12 – Total Solar Eclipses: 2027 – 2039

During the 13-year period 2027 through 2039, Earth will experience 10 total solar eclipses. Examination of the map in Figure 10 shows that only one of them is visible from North America (Alaska on 2033 Mar 30). In contrast, Australia is fortunate to have no less than four total solar eclipses during the same period (2028 July 22, 2030, Nov 25, 2037 July 13, and 2038 Dec 26). Information for each eclipse appears in *Table 12*.

Table 12 – Data for Total Solar Eclipses

Calendar Date	Time of Greatest Eclipse	Saros Number	Latitude	Longitude	Sun Alt	Path Width km	Duration Totality
2027 Aug 02	10:06:41	136	26°N	33°E	82°	258	06m23s
2028 Jul 22	02:55:29	146	16°S	127°E	53°	230	05m10s
2030 Nov 25	06:50:27	133	44°S	71°E	67°	169	03m44s
2031 Nov 14*	21:06:20	143	1°S	138°W	72°	38	01m08s
2033 Mar 30	18:01:25	120	71°N	156°W	11°	781	02m37s
2034 Mar 20	10:17:35	130	16°N	22°E	73°	159	04m09s
2035 Sep 02	01:55:36	145	29°N	158°E	68°	116	02m54s
2037 Jul 13	02:39:25	127	25°S	139°E	43°	201	03m58s
2038 Dec 26	00:58:58	142	40°S	164°E	73°	95	02m18s
2039 Dec 15	16:22:34	152	81°S	173°E	18°	380	01m51s

* The eclipse of 2031 Nov 14 is a Hybrid Eclipse or Annular/Total Eclipse.

Explanation of Column Headings in Table 12

Calendar Date – The UTC date of the eclipse at the instant of Greatest Eclipse
Time of Greatest Eclipse – The UTC Time of Greatest Eclipse
Saros Number – The number of the Saros series that the eclipse belongs to
Latitude – The latitude of the Moon's umbral shadow at instant of Greatest Eclipse
Longitude – The longitude of the Moon's umbral shadow at instant of Greatest Eclipse
Sun Alt – The altitude of the Sun at the instant of Greatest Eclipse
Path Width – The width of the path of totality (km) at the instant of Greatest Eclipse
Duration Totality – The central line duration of totality at the instant of Greatest Eclipse

Total Solar Eclipses from 2027 to 2039

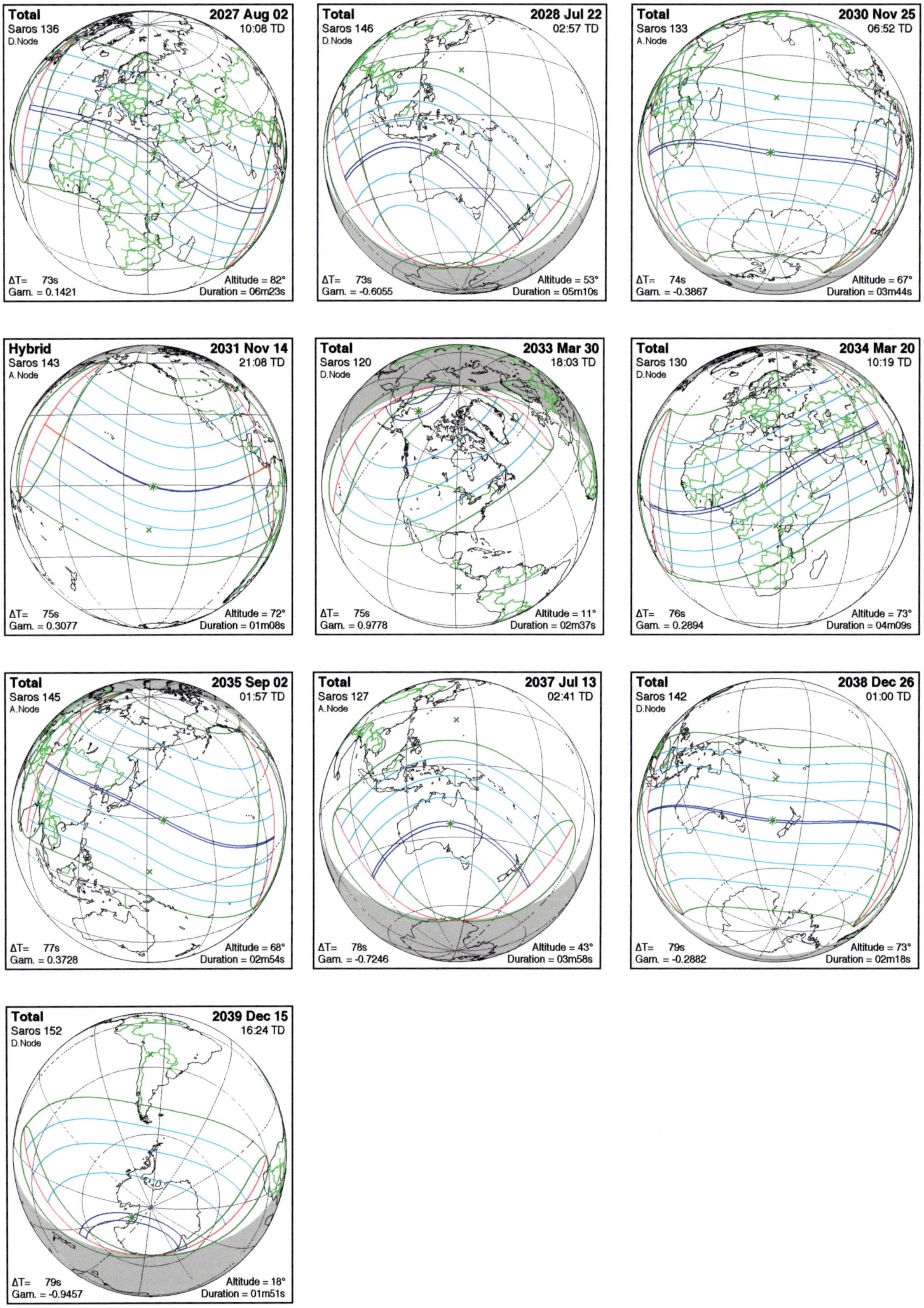

Figure 11 – Total Solar Eclipses from 2027 through 2039.
*Courtesy of **Thousand Year Canon of Solar Eclipses1501 to 2500**, F. Espenak, (2014)*

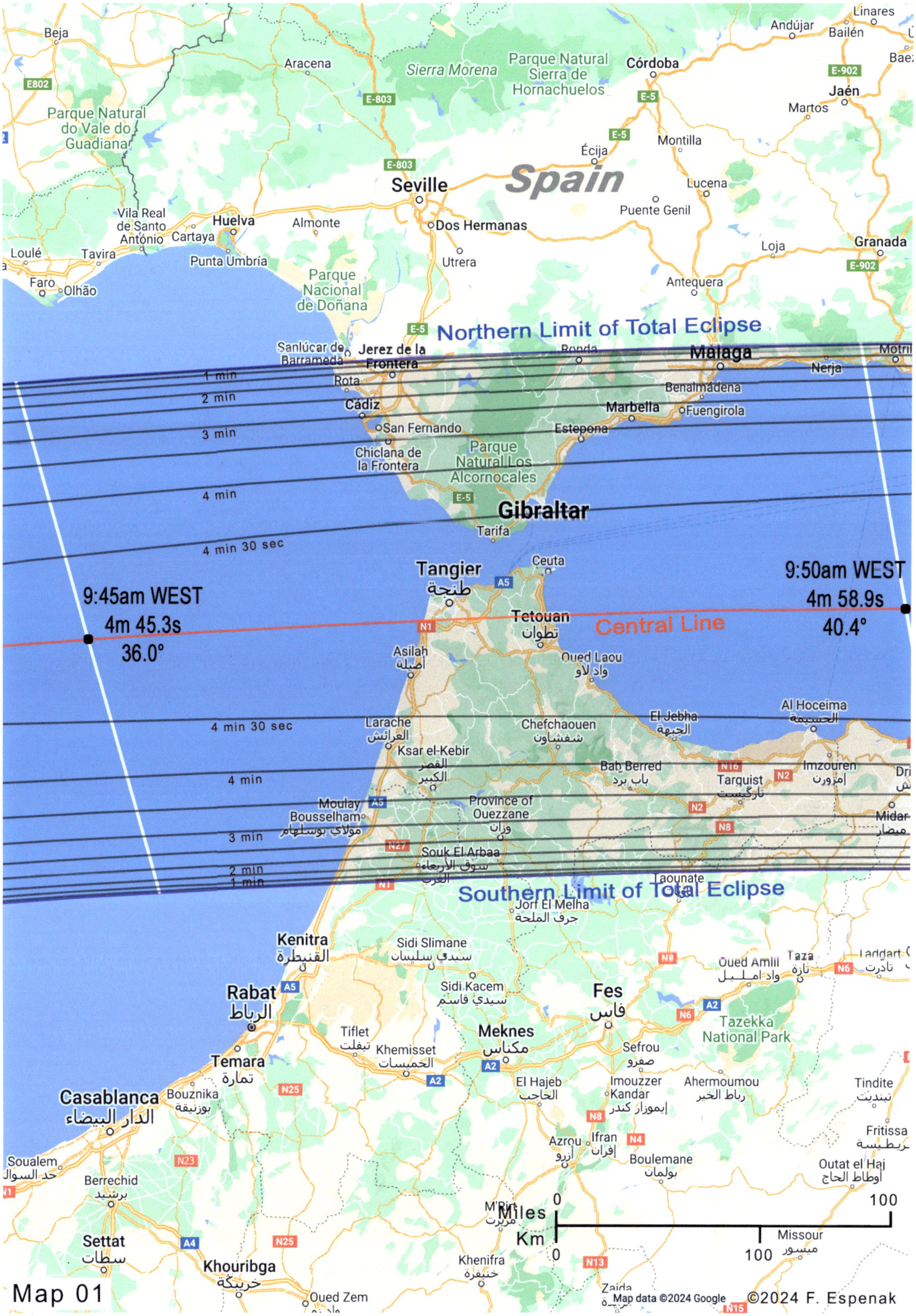
Spain
Seville
Córdoba
Granada
Málaga
Huelva
Jerez de la Frontera
Cádiz
Gibraltar
Tarifa
Ceuta
Tangier
Tetouan
Northern Limit of Total Eclipse
Southern Limit of Total Eclipse
Central Line
1 min
2 min
3 min
4 min
4 min 30 sec
9:45am WEST
4m 45.3s
36.0°
9:50am WEST
4m 58.9s
40.4°
Larache
Ksar el-Kebir
Chefchaouen
Al Hoceima
Kenitra
Rabat
Casablanca
Meknes
Fes
Miles
Km
0
100
Map 01
Map data ©2024 Google
©2024 F. Espenak

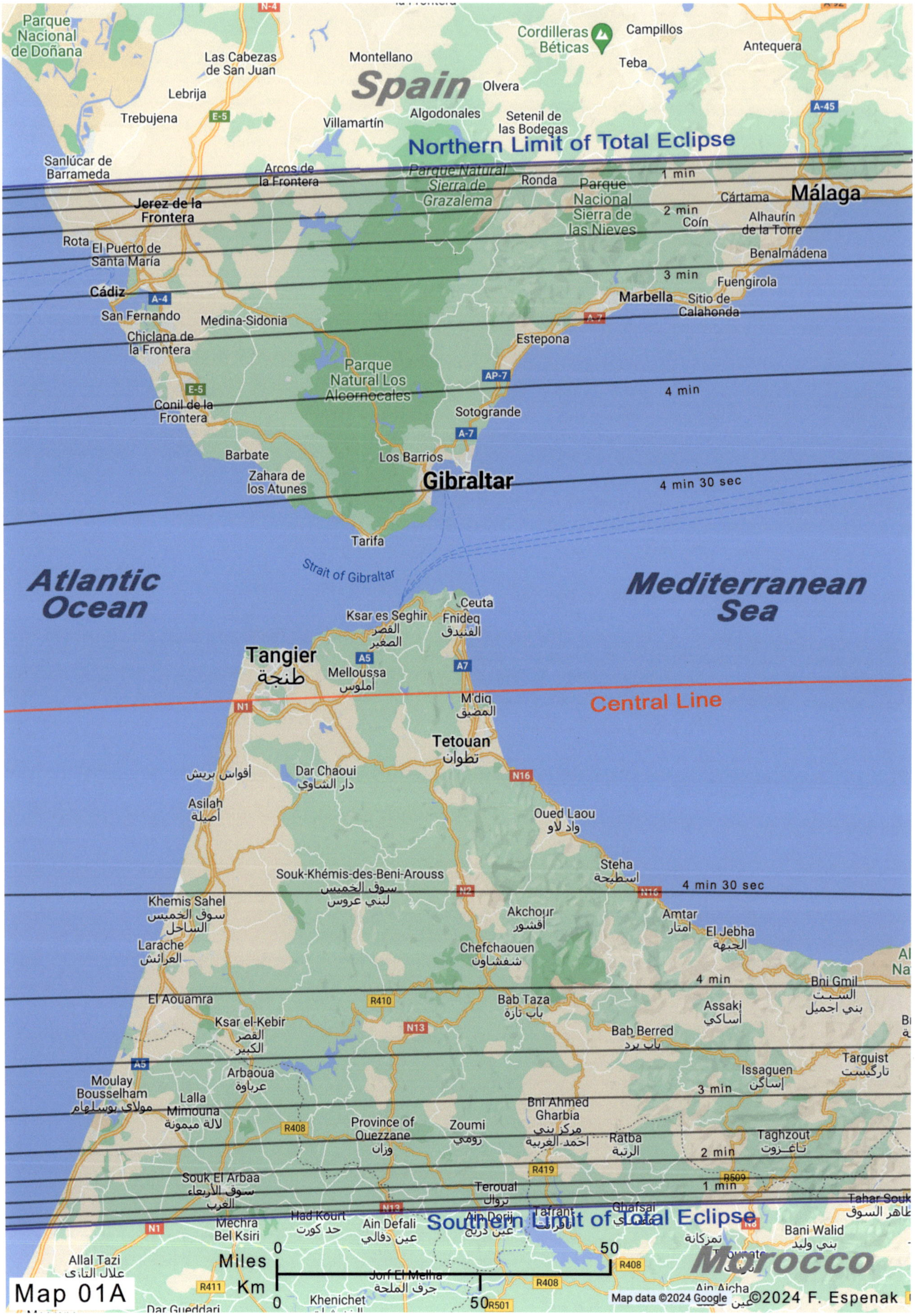
Spain
Northern Limit of Total Eclipse
Parque Nacional de Doñana
Las Cabezas de San Juan
Montellano
Cordilleras Béticas
Campillos
Antequera
Teba
Olvera
Lebrija
Trebujena
Villamartín
Algodonales
Setenil de las Bodegas
Sanlúcar de Barrameda
Arcos de la Frontera
Parque Natural Sierra de Grazalema
Ronda
Parque Nacional Sierra de las Nieves
1 min
Málaga
Cártama
Jerez de la Frontera
2 min
Coín
Alhaurín de la Torre
Rota
El Puerto de Santa María
Benalmádena
3 min
Fuengirola
Cádiz
Marbella
Sitio de Calahonda
San Fernando
Medina-Sidonia
Chiclana de la Frontera
Estepona
Parque Natural Los Alcornocales
4 min
Conil de la Frontera
Sotogrande
Barbate
Los Barrios
Zahara de los Atunes
Gibraltar
4 min 30 sec
Tarifa
Strait of Gibraltar
Atlantic Ocean
Mediterranean Sea
Ceuta
Ksar es Seghir
Fnideq
Tangier
طنجة
Melloussa
M'diq
Central Line
Tetouan
تطوان
Dar Chaoui
Asilah
Oued Laou
Steha
Souk-Khémis-des-Beni-Arouss
4 min 30 sec
Khemis Sahel
Akchour
Amtar
El Jebha
Larache
Chefchaouen
4 min
Bni Gmil
El Aouamra
Bab Taza
Assaki
Ksar el-Kebir
Bab Berred
Targuist
Moulay Bousselham
Arbaoua
Issaguen
3 min
Lalla Mimouna
Bni Ahmed Gharbia
Province of Ouezzane
Zoumi
Ratba
Taghzout
2 min
Souk El Arbaa
Teroual
1 min
Tahar Souk
Had Kourt
Ain Defali
Ain Dorij
Tafrant
Ghafsai
Southern Limit of Total Eclipse
Mechra Bel Ksiri
Bani Walid
Allal Tazi
Miles
Km
0
50
Morocco
Jorf El Melha
Ain Aicha
Map 01A
Dar Gueddari
Khenichet
Map data ©2024 Google
©2024 F. Espenak

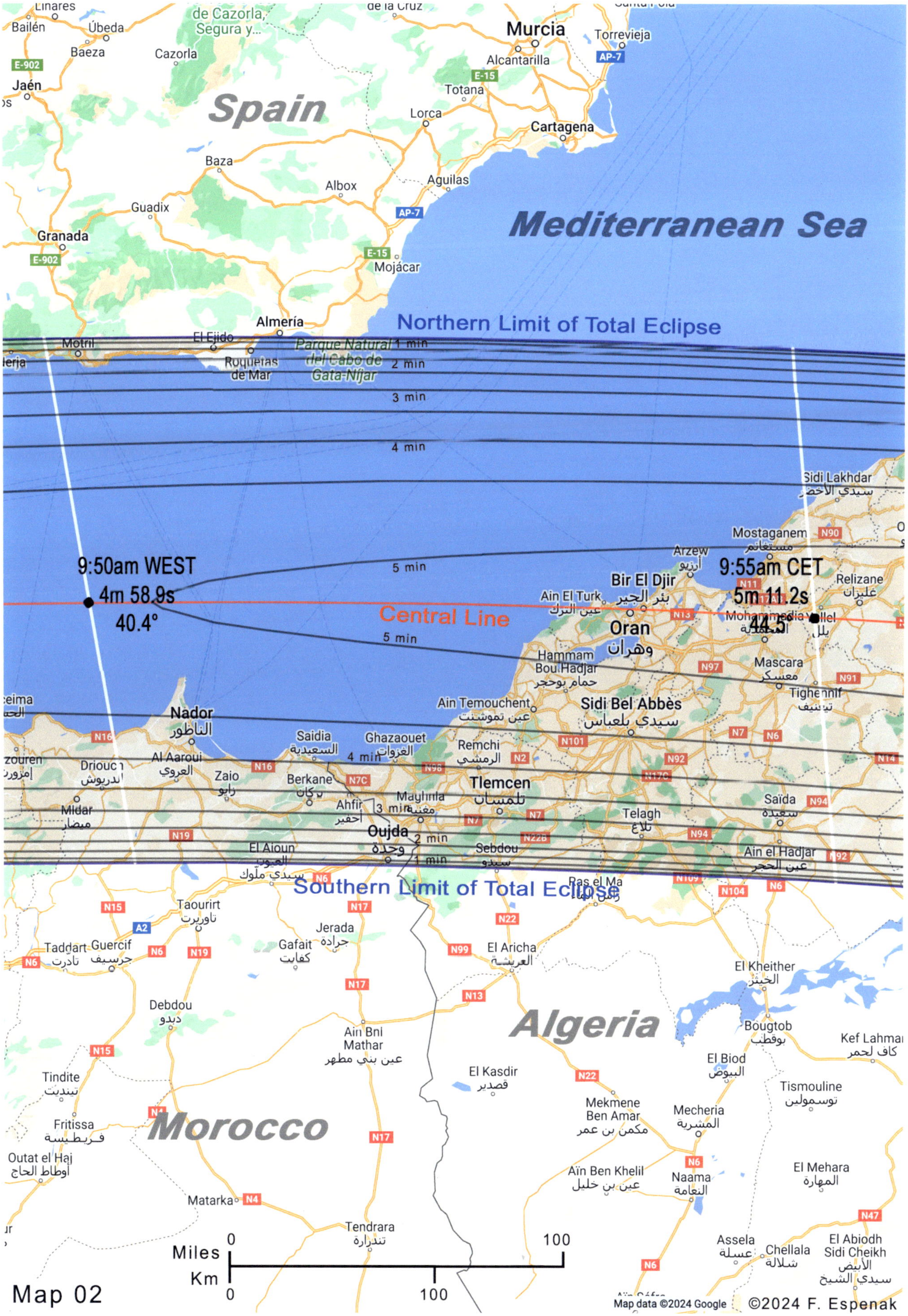
Spain
Mediterranean Sea
Algeria
Morocco
Northern Limit of Total Eclipse
Southern Limit of Total Eclipse
Central Line
1 min
2 min
3 min
4 min
5 min
5 min
4 min
3 min
2 min
1 min
9:50am WEST
4m 58.9s
40.4°
9:55am CET
5m 11.2s
44.5°
Murcia
Torrevieja
Alcantarilla
Totana
Lorca
Cartagena
Granada
Guadix
Baza
Albox
Aguilas
Mojácar
Almería
Motril
El Ejido
Roquetas de Mar
Parque Natural del Cabo de Gata-Níjar
Baena
Úbeda
Bailén
Jaén
Cazorla
Nador
Saidia
Ghazaouet
Berkane
Oujda
Tlemcen
Oran
Sidi Bel Abbès
Mostaganem
Relizane
Mascara
Arzew
Bir El Djir
Ain El Turk
Ain Temouchent
Remchi
Maghnia
Sebdou
Telagh
Saïda
Tighennif
Ain el Hadjar
Sidi Lakhdar
Hammam Bou Hadjar
Al Aaroui
Zaio
Driouch
Midar
Ahfir
El Aioun
Taourirt
Guercif
Taddart
Jerada
Gafait
El Aricha
Debdou
Ain Bni Mathar
El Kasdir
Mekmene Ben Amar
Mecheria
Ain Ben Khelil
Naama
El Kheither
Bougtob
El Biod
Kef Lahmar
Tismouline
El Mehara
Tindite
Fritissa
Outat el Haj
Matarka
Tendrara
Assela
Chellala
El Abiodh Sidi Cheikh
Miles 0 100
Km 0 100
Map 02
Map data ©2024 Google
©2024 F. Espenak

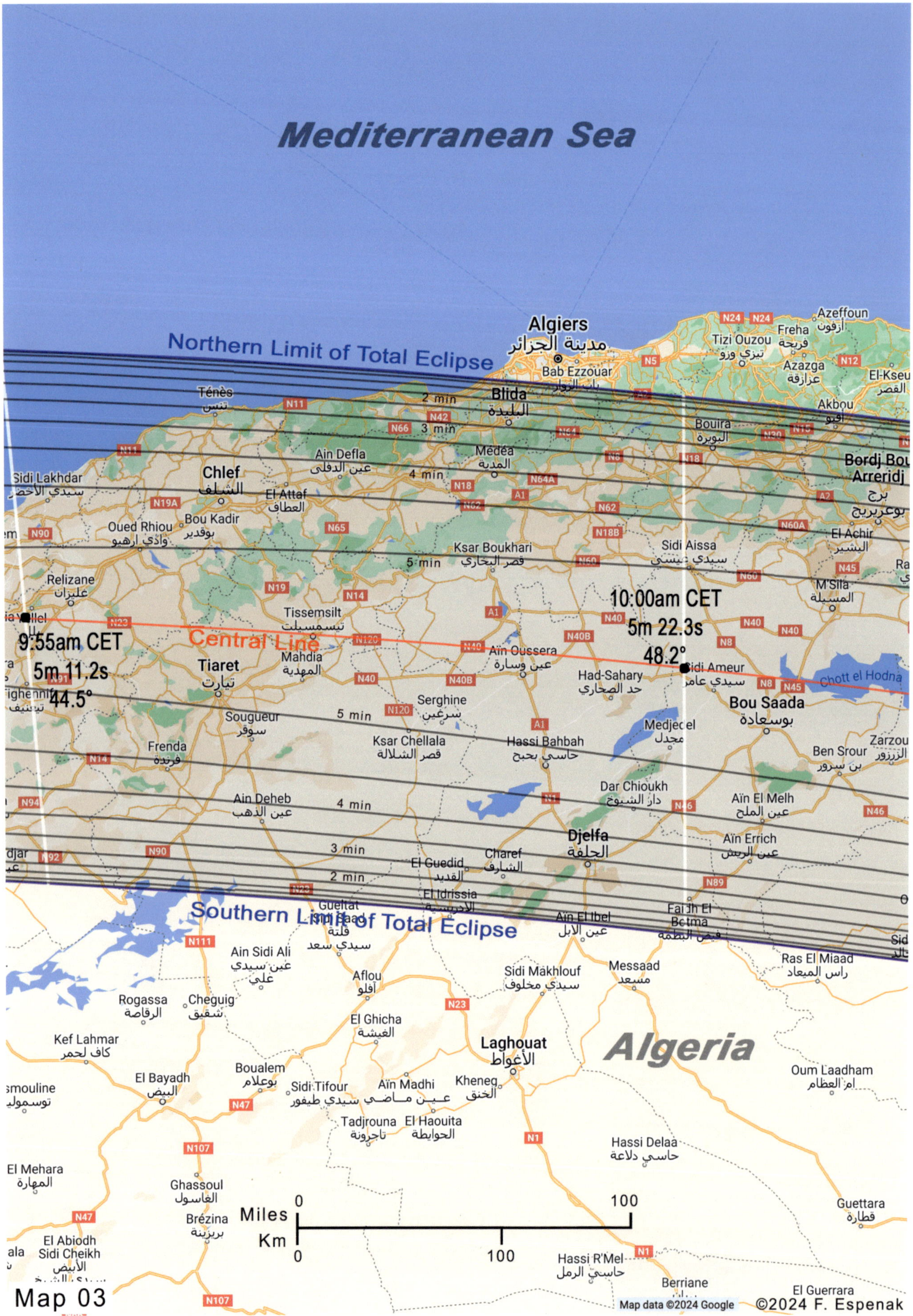
Mediterranean Sea
Northern Limit of Total Eclipse
Southern Limit of Total Eclipse
Central Line
9:55am CET
5m 11.2s
44.5°
10:00am CET
5m 22.3s
48.2°
2 min
3 min
4 min
5 min
Algiers
Blida
Médéa
Chlef
Tiaret
Djelfa
Bou Saada
Laghouat
Algeria
Miles
Km
0
100
Map data ©2024 Google
©2024 F. Espenak

Map 03

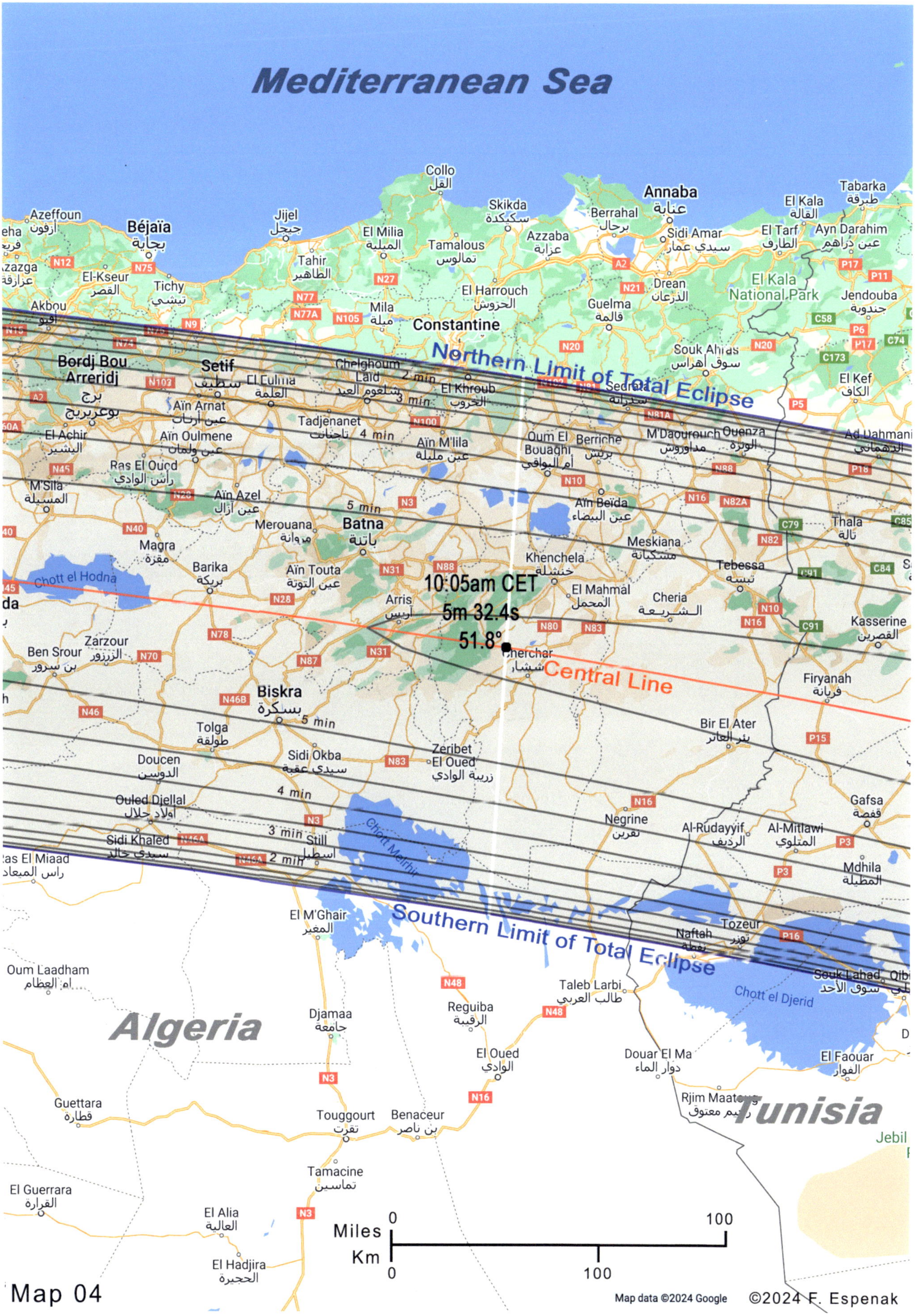
Mediterranean Sea
Northern Limit of Total Eclipse
Central Line
Southern Limit of Total Eclipse
10:05am CET
5m 32.4s
51.8°
2 min
3 min
4 min
5 min
Algeria
Tunisia
Mediterranean Sea
Collo
Skikda
Annaba
Tabarka
El Kala
Jijel
Béjaïa
Azeffoun
El Milia
Tamalous
Azzaba
Berrahal
Sidi Amar
El Tarf
Ayn Darahim
Tahir
El-Kseur
Tichy
El Harrouch
Drean
El Kala National Park
Jendouba
Akbou
Mila
Constantine
Guelma
Souk Ahras
Bordj Bou Arreridj
Setif
El Eulma
Chelghoum Laid
El Khroub
Sedrata
El Kef
Aïn Arnat
Aïn Oulmene
Tadjenanet
Aïn M'lila
Oum El Bouaghi
Berriche
M'Daourouch
Ouenza
Ad Dahmani
El Achir
Ras El Oued
M'Sila
Aïn Azel
Merouana
Batna
Aïn Beïda
Thala
Magra
Barika
Meskiana
Khenchela
Tebessa
Chott el Hodna
Aïn Touta
Arris
El Mahmal
Cheria
Kasserine
Zarzour
Ben Srour
Biskra
Chechar
Firyanah
Tolga
Sidi Okba
Zeribet El Oued
Bir El Ater
Doucen
Ouled Djellal
Sidi Khaled
Still
Negrine
Al-Rudayyif
Al-Mitlawi
Gafsa
Mdhila
Chott Melrhir
El M'Ghair
Naftah
Tozeur
Souk Lahad
Oum Laadham
Taleb Larbi
Chott el Djerid
Reguiba
Djamaa
El Oued
Douar El Ma
El Faouar
Guettara
Rjim Maatoug
Touggourt
Benaceur
Jebil
Tamacine
El Guerrara
El Alia
El Hadjira
Miles
Km
0
100
Map 04
Map data ©2024 Google
©2024 F. Espenak

Mediterranean Sea
Tunisia
Bizerte
Ras Jebel
Tabarka
Mateur
Tunis
Ayn Darahim
Beja
Ben Arous
Kelibia
Pantelleria
Tracino
Scauri
Jendouba
Zaghouan
Hammamet
Yasmine Hammamet
El Kef
Siliana
Ad Dahmani
Maktar
Sousse
Kairouan
Monastir
Linosa
Thala
Northern Limit of Total Eclipse
Lampedusa
Hajeb El Ayoun
Sabibah
Mahdia
El Jem
Chebba
Kasserine
Subaytilah
Sidi Bouzid
Jebiniana
Firyanah
Bir Al Huffay
Regueb
Bir Ali Ben Khlifa
Sfax
Sidi Ali Ben Aoun
Agareb
Thyna
Al Mazzunah
Sened
Gafsa
Al-Maharas
Skhira
Mdhila
2 min
3 min
4 min
5 min
10:10am CET
5m 41.5s
55.2°
Central Line
10:15am CET
5m 49.6s
58.4°
Al Hammah
Gabes
Houmt Souk
Djerba Midun
Djerba
Souk Lahad
Qibili
Mareth
Djerba Ajim
Zarzis
Southern Limit of Total Eclipse
Douz
El Faouar
Medenine
Ben Gardane
Ra's Ajdir
Zelten
Zuwara
Jebil National Park
Tataouine
Samar
Al Assah
Ksar Ghilane
Tunisia
Ksar Ouled Debbab
Sabratah
Surman
Az-Zāwiyah
Janzur
Al-'Azīzīya
Libya
Garyan
Miles
Km
0
100
Remada
Map 05
Map data ©2024 Google
©2024 F. Espenak

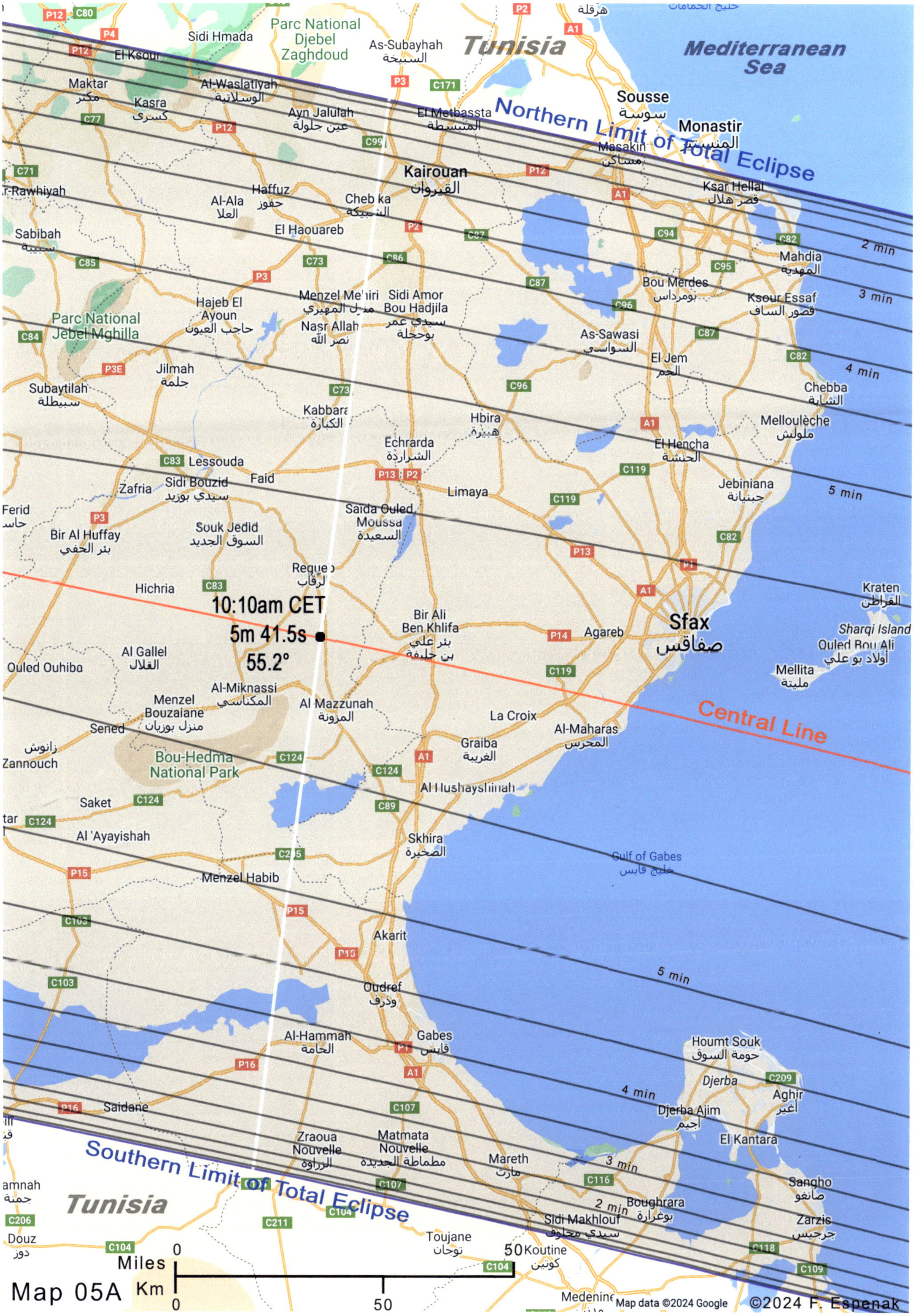
Tunisia
Mediterranean Sea
Northern Limit of Total Eclipse
Central Line
Southern Limit of Total Eclipse
10:10am CET
5m 41.5s
55.2°
2 min
3 min
4 min
5 min
5 min
4 min
3 min
2 min
Parc National Djebel Zaghdoud
Sidi Hmada
El Ksour
Maktar
Kasra
Al-Waslatiyah
As-Subayhah
Sousse
Monastir
Masakin
Ayn Jalulah
El Metbassta
Kairouan
Ksar Hellal
Mahdia
Haffuz
Al-Ala
Cheb ka
El Haouareb
Sabibah
Bou Merdes
Ksour Essaf
Parc National Jebel Mghilla
Hajeb El Ayoun
Menzel Mehiri
Sidi Amor Bou Hadjila
Nasr Allah
As-Sawasi
El Jem
Subaytilah
Jilmah
Kabbara
Hbira
Chebba
Melloulèche
Echrarda
El Hencha
Lessouda
Sidi Bouzid
Zafria
Faid
Limaya
Jebiniana
Ferid
Saida Ouled Moussa
Bir Al Huffay
Souk Jedid
Hichria
Sfax
Agareb
Bir Ali Ben Khlifa
Kraten
Sharqi Island
Ouled Bou Ali
Mellita
Ouled Ouhiba
Al Gallel
Al-Miknassi
Menzel Bouzaiane
Al Mazzunah
La Croix
Al-Maharas
Graiba
Sened
Zannouch
Bou-Hedma National Park
Saket
Al 'Ayayishah
Skhira
Gulf of Gabes
Menzel Habib
Akarit
Oudref
Al-Hammah
Gabes
Houmt Souk
Djerba
Aghir
Djerba Ajim
El Kantara
Saidane
Zraoua Nouvelle
Matmata Nouvelle
Mareth
Sangho
Boughrara
Zarzis
Sidi Makhlouf
Toujane
Koutine
Douz
Medenine
Miles
Km
0
50
0
50
Map 05A
Map data ©2024 Google
©2024 F. Espenak

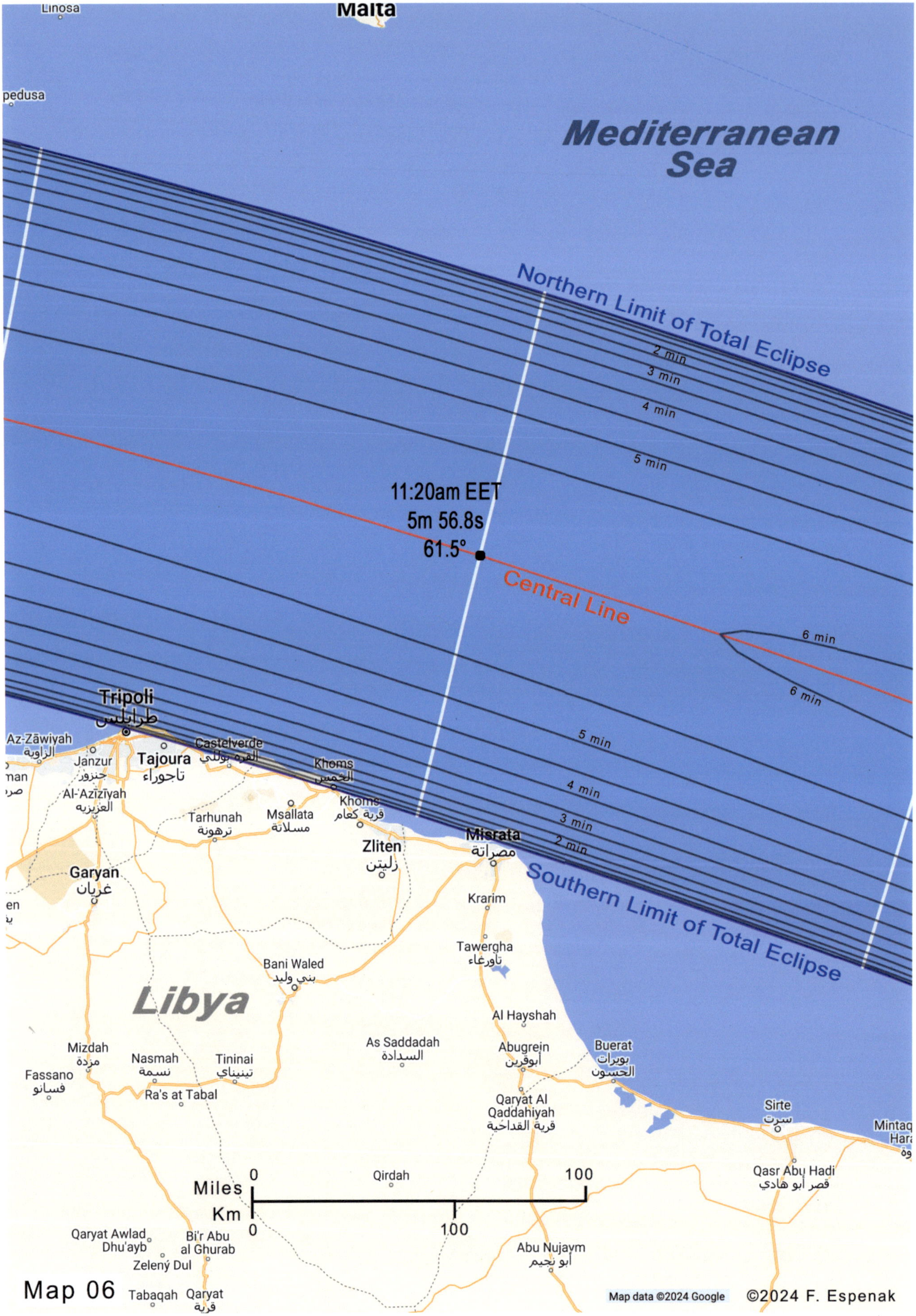
Malta
Linosa
pedusa
Mediterranean Sea
Northern Limit of Total Eclipse
2 min
3 min
4 min
5 min
11:20am EET
5m 56.8s
61.5°
Central Line
6 min
6 min
5 min
4 min
3 min
2 min
Southern Limit of Total Eclipse
Tripoli
Az-Zāwiyah
Janzur
Tajoura
Castelverde
Khoms
Al-'Azīzīyah
Tarhunah
Msallata
Khoms
Zliten
Misrata
Garyan
Krarim
Tawergha
Bani Waled
Libya
Al Hayshah
As Saddadah
Abugrein
Buerat
Mizdah
Nasmah
Tininai
Fassano
Ra's at Tabal
Qaryat Al Qaddahiyah
Sirte
Qasr Abu Hadi
Qirdah
Miles
Km
0
100
0
100
Qaryat Awlad Dhu'ayb
Bi'r Abu al Ghurab
Zelený Dul
Abu Nujaym
Tabaqah
Qaryat
Map 06
Map data ©2024 Google
©2024 F. Espenak

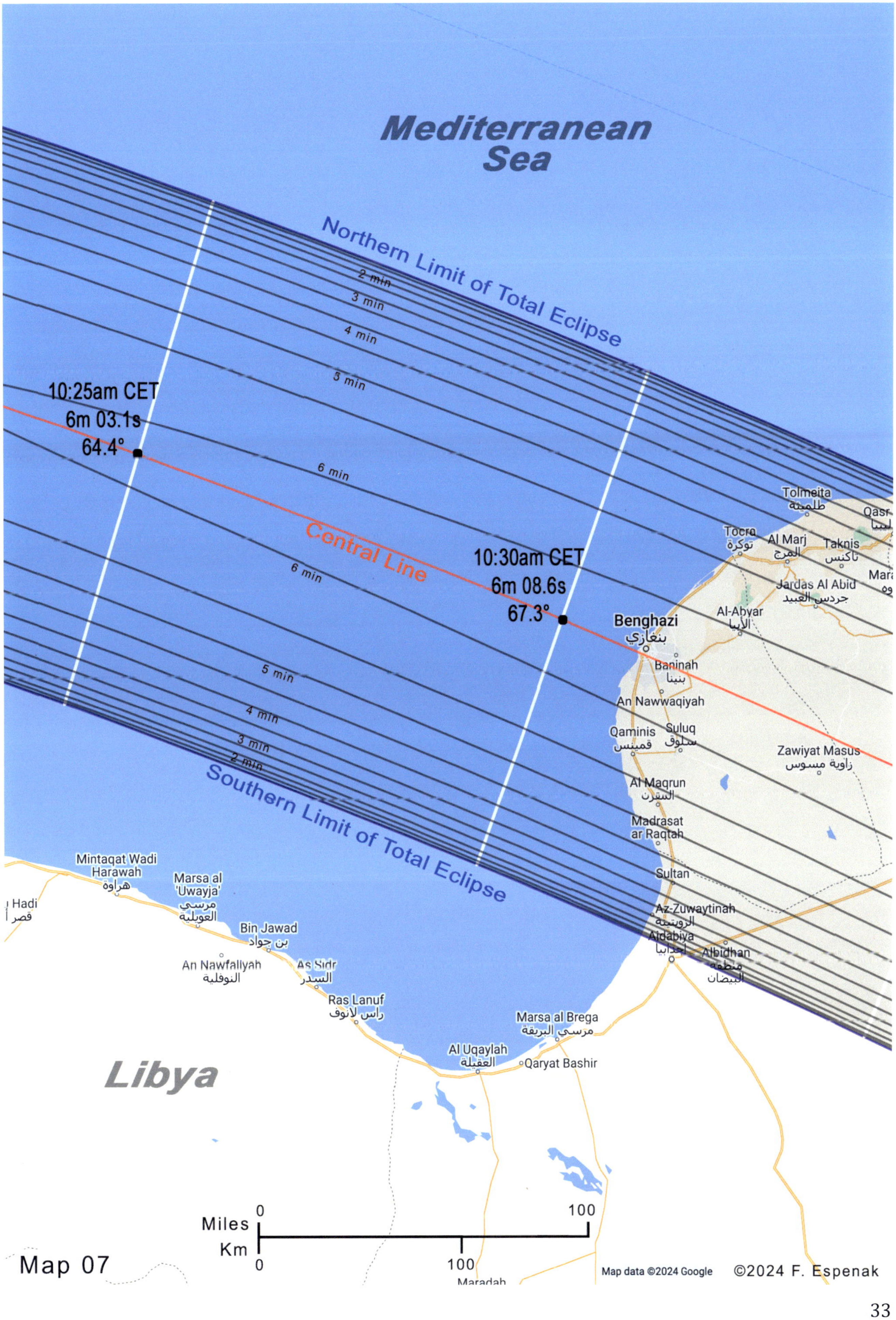
Mediterranean Sea
Northern Limit of Total Eclipse
2 min
3 min
4 min
5 min
6 min
10:25am CET
6m 03.1s
64.4°
Central Line
6 min
10:30am CET
6m 08.6s
67.3°
5 min
4 min
3 min
2 min
Southern Limit of Total Eclipse
Tolmeita
Tocra
Al Marj
Taknis
Jardas Al Abid
Benghazi
Al-Abyar
Baninah
An Nawwaqiyah
Qaminis
Suluq
Zawiyat Masus
Al Maqrun
Madrasat ar Raqtah
Sultan
Az-Zuwaytinah
Ajdabiya
Albidhan
Mintaqat Wadi Harawah
Marsa al 'Uwayja'
Bin Jawad
An Nawfaliyah
As Sidr
Ras Lanuf
Marsa al Brega
Al Uqaylah
Qaryat Bashir
Libya
Miles 0 100
Km 0 100
Map 07
Map data ©2024 Google
©2024 F. Espenak

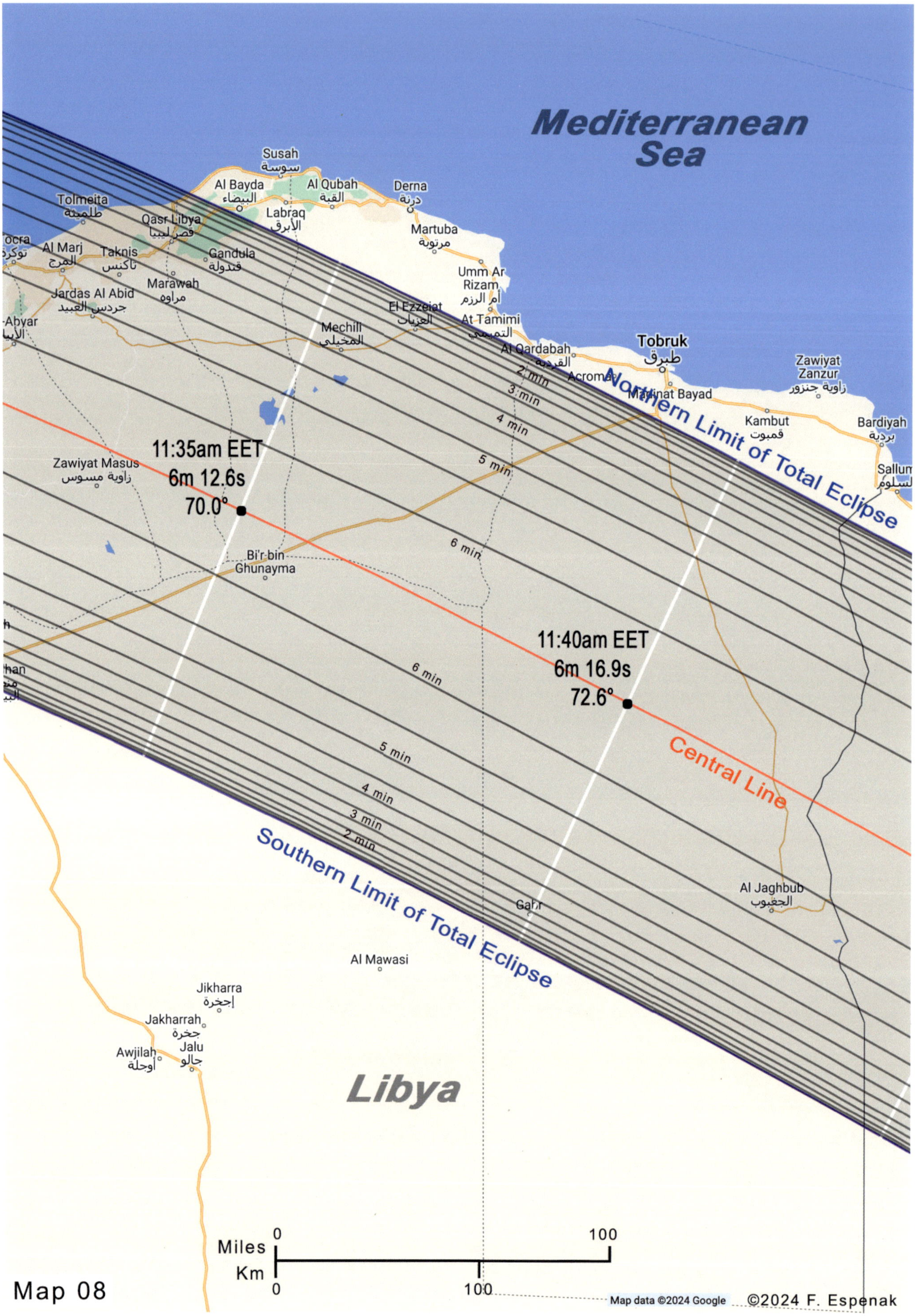
Mediterranean Sea
Susah
سوسة
Al Bayda
البيضاء
Al Qubah
القبة
Derna
درنة
Tolmeita
طلميثة
Qasr Libya
قصر ليبيا
Labraq
الأبرق
Martuba
مرتوبة
Al Marj
المرج
Taknis
تاكنس
Gandula
قندولة
Marawah
مراوه
Jardas Al Abid
جردس العبيد
Umm Ar Rizam
أم الرزم
El Ezzeiat
العزيات
At Tamimi
التميمي
Mechili
المخيلي
Al Qardabah
القردبة
Tobruk
طبرق
Acroma
Madinat Bayad
Zawiyat Zanzur
زاوية جنزور
Kambut
قمبوت
Bardiyah
بردية
Northern Limit of Total Eclipse
2 min
3 min
4 min
5 min
6 min
11:35am EET
6m 12.6s
70.0°
Zawiyat Masus
زاوية مسوس
Bi'r bin Ghunayma
11:40am EET
6m 16.9s
72.6°
Central Line
6 min
5 min
4 min
3 min
2 min
Southern Limit of Total Eclipse
Al Jaghbub
الجغبوب
Gabr
Al Mawasi
Jikharra
إجخرة
Jakharrah
جخرة
Jalu
جالو
Awjilah
أوجلة
Libya
Miles 0 100
Km 0 100
Map 08
Map data ©2024 Google
©2024 F. Espenak

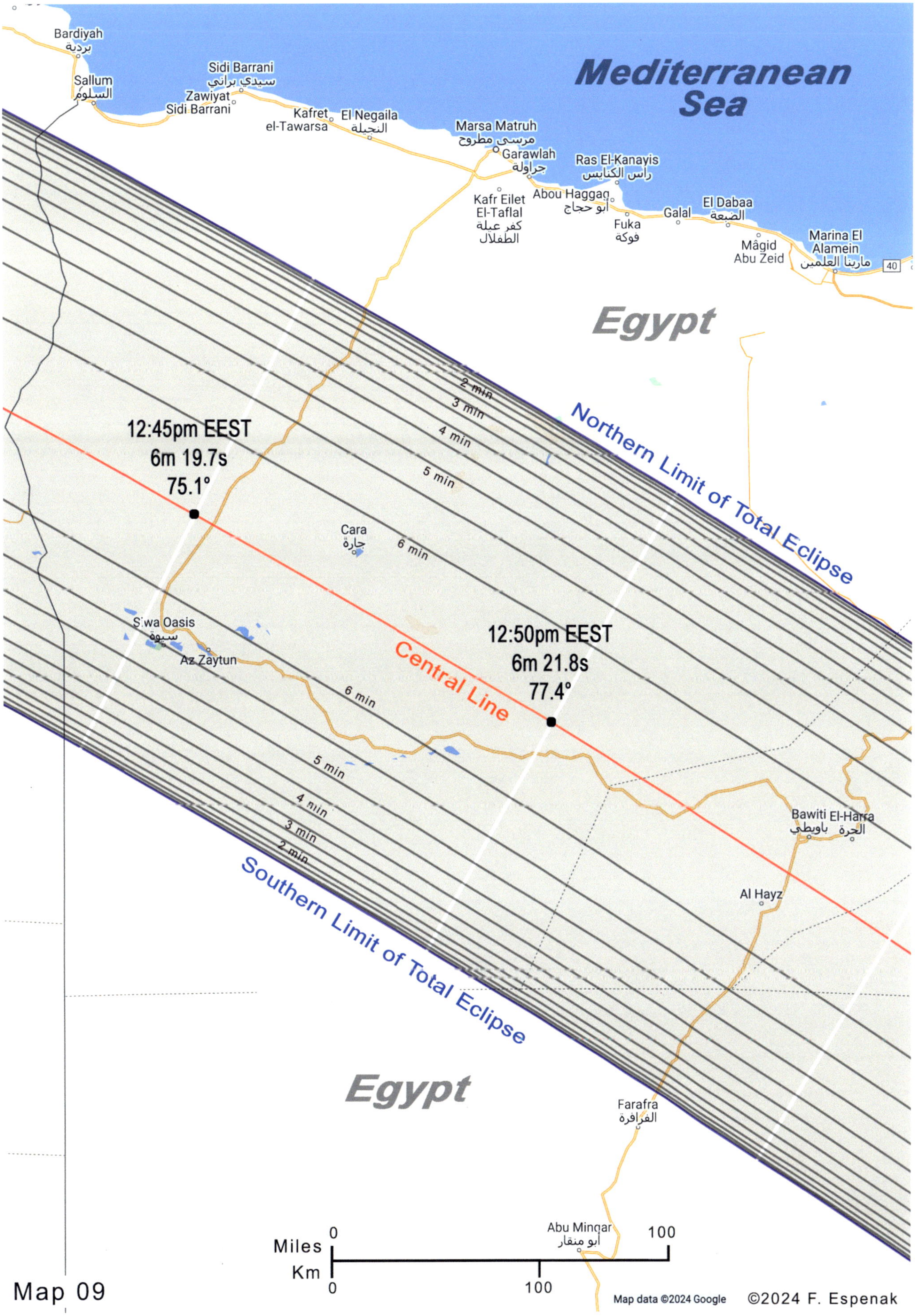
Mediterranean Sea
Bardiyah
بردية
Sallum
السلوم
Sidi Barrani
سيدي براني
Zawiyat Sidi Barrani
Kafret el-Tawarsa
El Negaila
النجيلة
Marsa Matruh
مرسى مطروح
Garawlah
جراولة
Ras El-Kanayis
راس الكنايس
Kafr Eilet El-Taflal
كفر عيلة الطفلال
Abou Haggag
أبو حجاج
Fuka
فوكة
Galal
El Dabaa
الضبعة
Mâgid Abu Zeid
Marina El Alamein
مارينا العلمين
40
Egypt
2 min
3 min
4 min
5 min
6 min
Northern Limit of Total Eclipse
12:45pm EEST
6m 19.7s
75.1°
Cara
جارة
Siwa Oasis
سيوة
Az Zaytun
Central Line
12:50pm EEST
6m 21.8s
77.4°
6 min
5 min
4 min
3 min
2 min
Bawiti
باويطي
El-Harra
الحرة
Al Hayz
Southern Limit of Total Eclipse
Egypt
Farafra
الفرافرة
Abu Minqar
أبو منقار
Miles 0 100
Km 0 100
Map data ©2024 Google
©2024 F. Espenak

Map 09

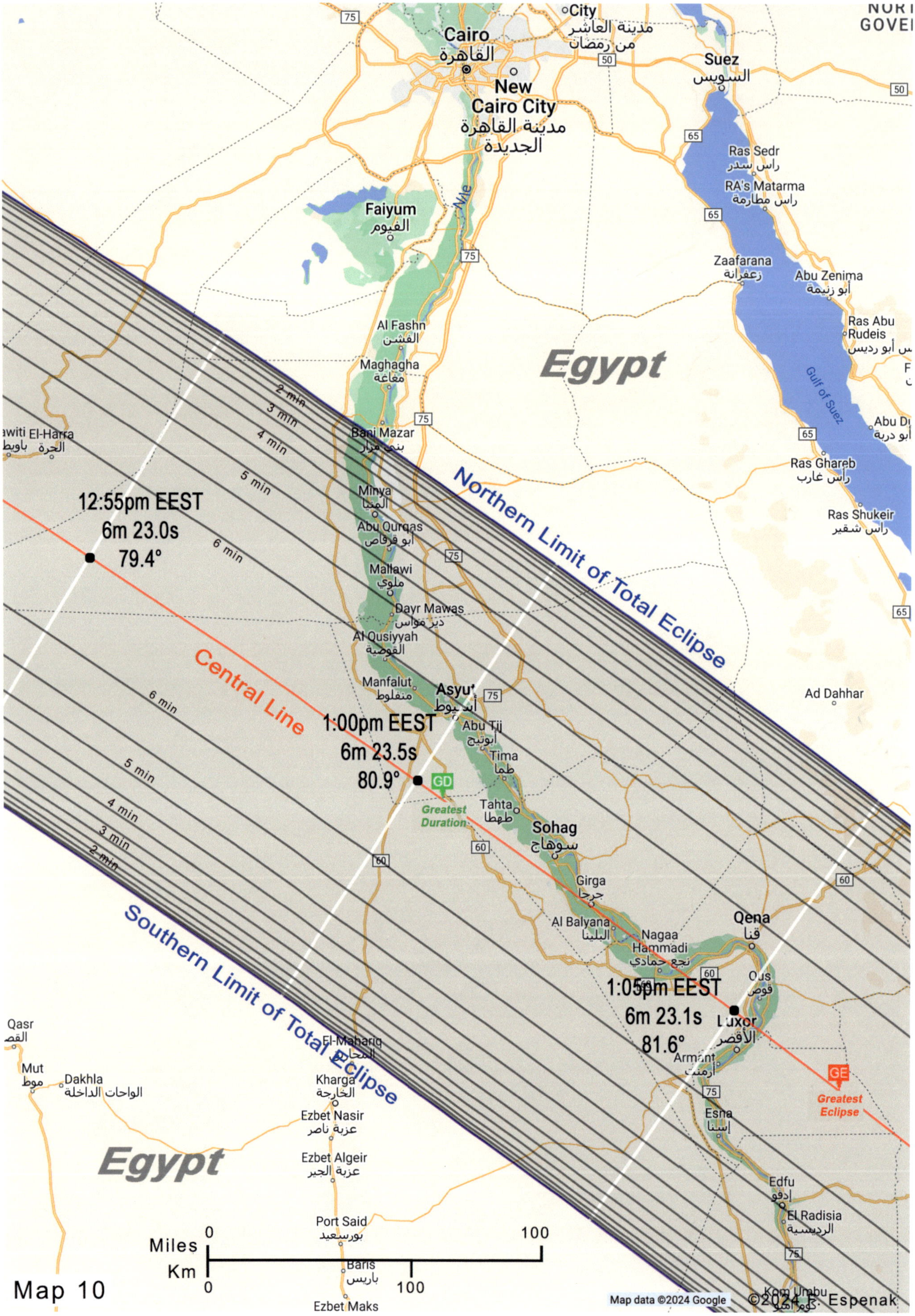
Cairo
New Cairo City
Suez
Ras Sedr
RA's Matarma
Faiyum
Nile
Zaafarana
Abu Zenima
Ras Abu Rudeis
Gulf of Suez
Al Fashn
Maghagha
Egypt
Bani Mazar
El-Harra
Ras Ghareb
Ras Shukeir
Minya
Abu Qurqas
Mallawi
Dayr Mawas
Al Qusiyyah
Manfalut
Asyut
Abu Tij
Tima
Tahta
Sohag
Girga
Al Balyana
Nagaa Hammadi
Qena
Qus
Luxor
Armant
Esna
Edfu
El Radisia
Kom Ombu
Ad Dahhar
Qasr
Mut
Dakhla
El-Mahariq
Kharga
Ezbet Nasir
Ezbet Algeir
Port Said
Baris
Ezbet Maks
Northern Limit of Total Eclipse
Southern Limit of Total Eclipse
Central Line
2 min
3 min
4 min
5 min
6 min
12:55pm EEST
6m 23.0s
79.4°
1:00pm EEST
6m 23.5s
80.9°
1:05pm EEST
6m 23.1s
81.6°
GD
Greatest Duration
GE
Greatest Eclipse
Miles 0 100
Km 0 100
Map 10
Map data ©2024 Google
©2024 F. Espenak

Egypt
Northern Limit of Total Eclipse
Central Line
Miles 0 50
Km 0 50
1:00pm EEST
6m 23.5s
80.9°
1:05pm EEST
6m 23.1s
81.6°
GD
Greatest Duration
2 min
3 min
4 min
5 min
6 min
Bani Mazar
Matay
Umm as Sas
Samalut
Al Birjayah
Minya
Abu Qurqas
Mallawi
Dayr Mawas
Dayrout
Al Qusiyyah
Manfalut
Abnub
Asyut
Abu Tij
Tima
Tahta
Juhaynah
Saqultah
Sohag
Al Minshah
Girga
Al Balyana
Dishna
Qena
Nagaa Hammadi
Qift
Qus
Al Ash Shi
Luxor
Map 10A
Map data ©2024 Google
©2024 F. Espenak

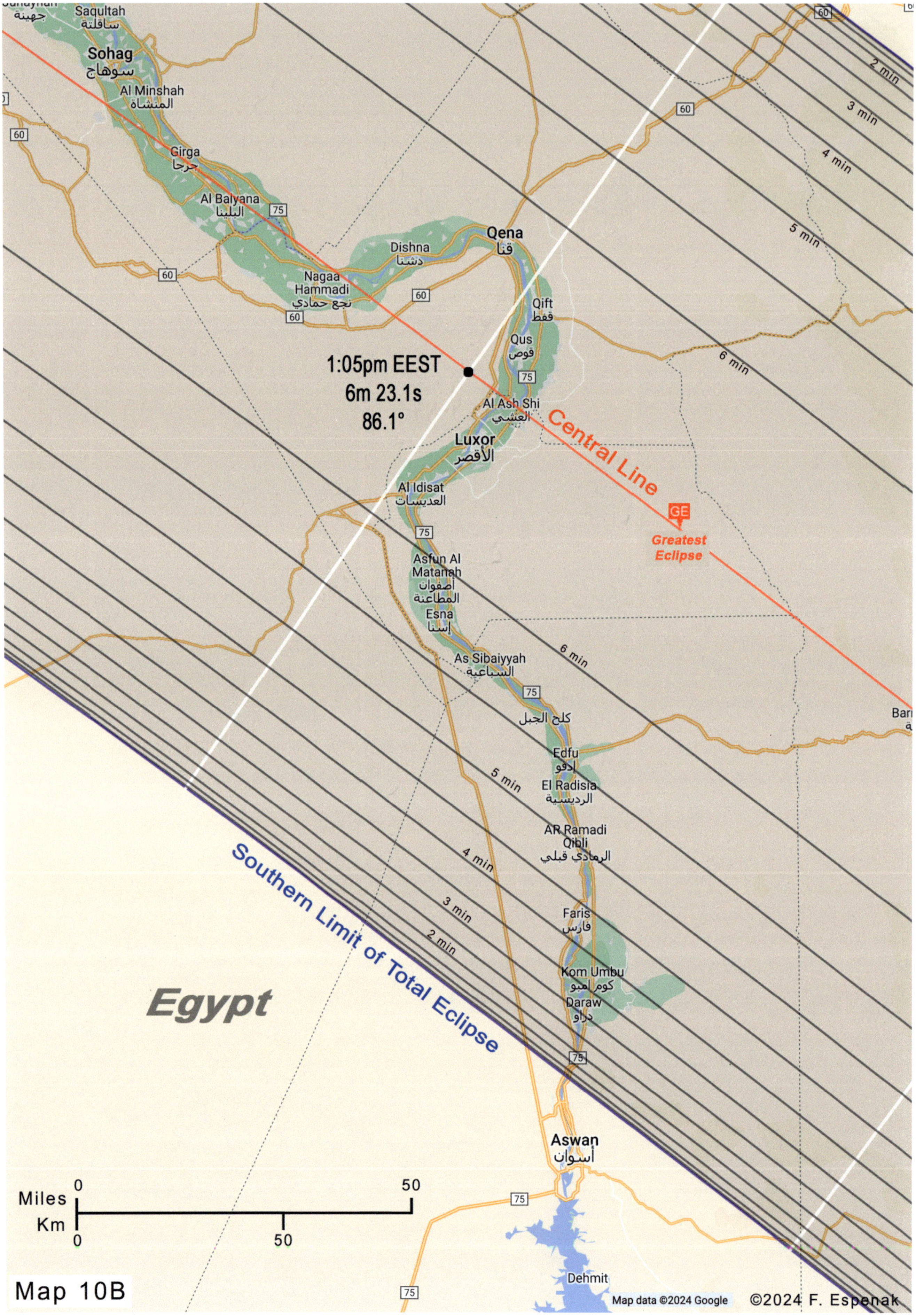

Saqultah
Sohag
Al Minshah
Girga
Al Balyana
Dishna
Qena
Nagaa Hammadi
Qift
Qus
1:05pm EEST
6m 23.1s
86.1°
Al Ash Shi
Luxor
Central Line
Al Idisat
GE
Greatest Eclipse
Asfun Al Matanah
Esna
As Sibaiyyah
Edfu
El Radisia
AR Ramadi Qibli
Faris
Kom Umbu
Daraw
Aswan
Dehmit
2 min
3 min
4 min
5 min
6 min
Southern Limit of Total Eclipse
Egypt
Miles 0 50
Km 0 50
Map 10B
Map data ©2024 Google
©2024 F. Espenak

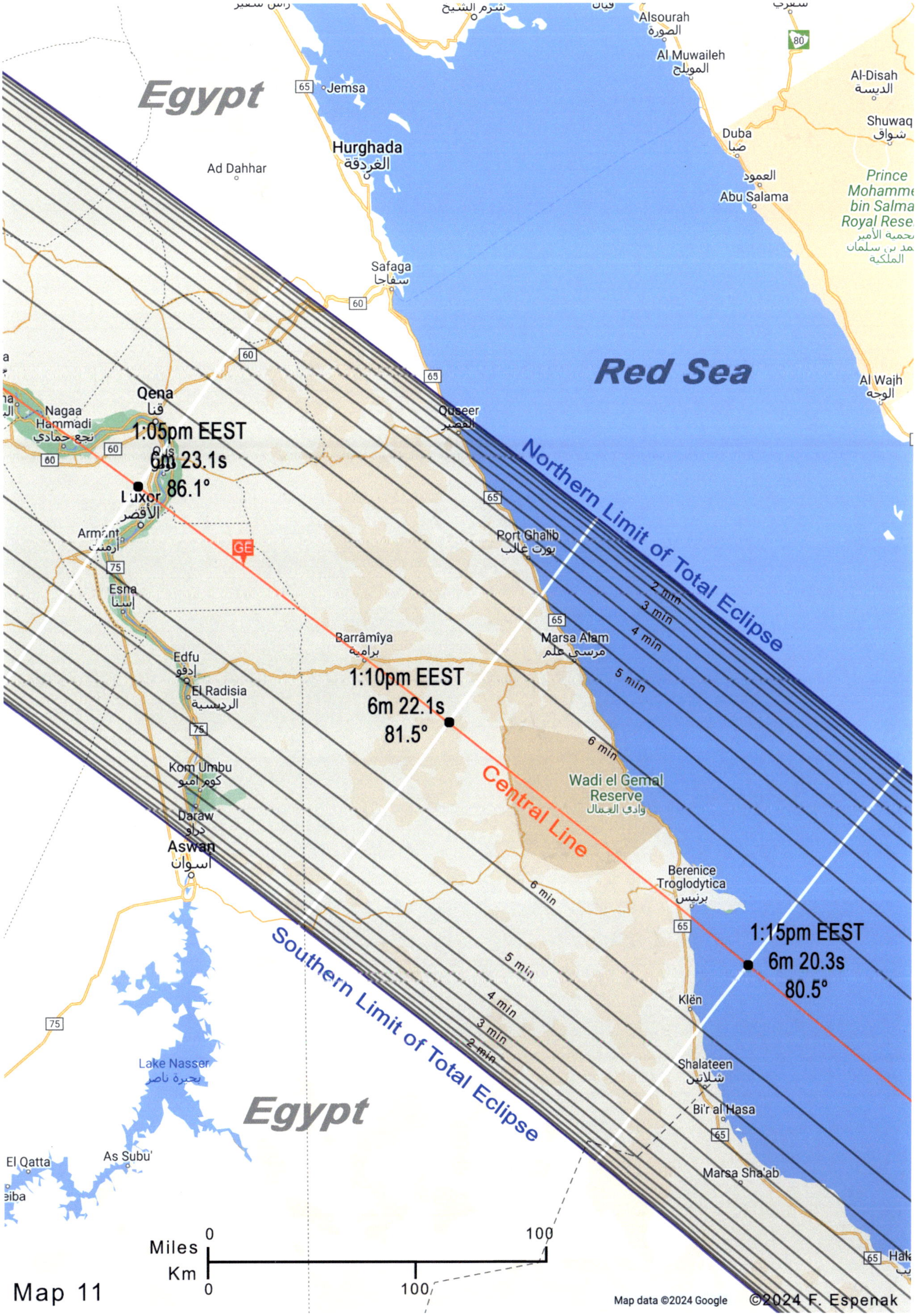
Egypt
Red Sea
Jemsa
Hurghada
الغردقة
Ad Dahhar
Alsourah
الصورة
Al Muwaileh
المويلح
Al-Disah
الديسة
Shuwaq
شواق
Duba
ضبا
العمود
Abu Salama
Prince
Mohammed
bin Salman
Royal Reserve
Safaga
سفاجا
Al Wajh
الوجه
Qena
قنا
Nagaa
Hammadi
نجع حمادي
Quseer
القصير
1:05pm EEST
6m 23.1s
86.1°
Luxor
الأقصر
Armant
Esna
إسنا
Port Ghalib
بورت غالب
Northern Limit of Total Eclipse
2 min
3 min
4 min
5 min
6 min
Barramiya
برامية
Marsa Alam
مرسى علم
Edfu
إدفو
El Radisia
الرديسية
1:10pm EEST
6m 22.1s
81.5°
Kom Umbu
كوم امبو
Wadi el Gemal
Reserve
وادي الجمال
Central Line
Daraw
دراو
Aswan
أسوان
Berenice
Troglodytica
برنيس
6 min
1:15pm EEST
6m 20.3s
80.5°
Southern Limit of Total Eclipse
5 min
4 min
3 min
2 min
Klën
Shalateen
شلاتين
Lake Nasser
بحيرة ناصر
Egypt
Bi'r al Hasa
El Qatta
As Subu'
Marsa Sha'ab
Miles
Km
0
100
0
100
Map 11
Map data ©2024 Google
©2024 F. Espenak

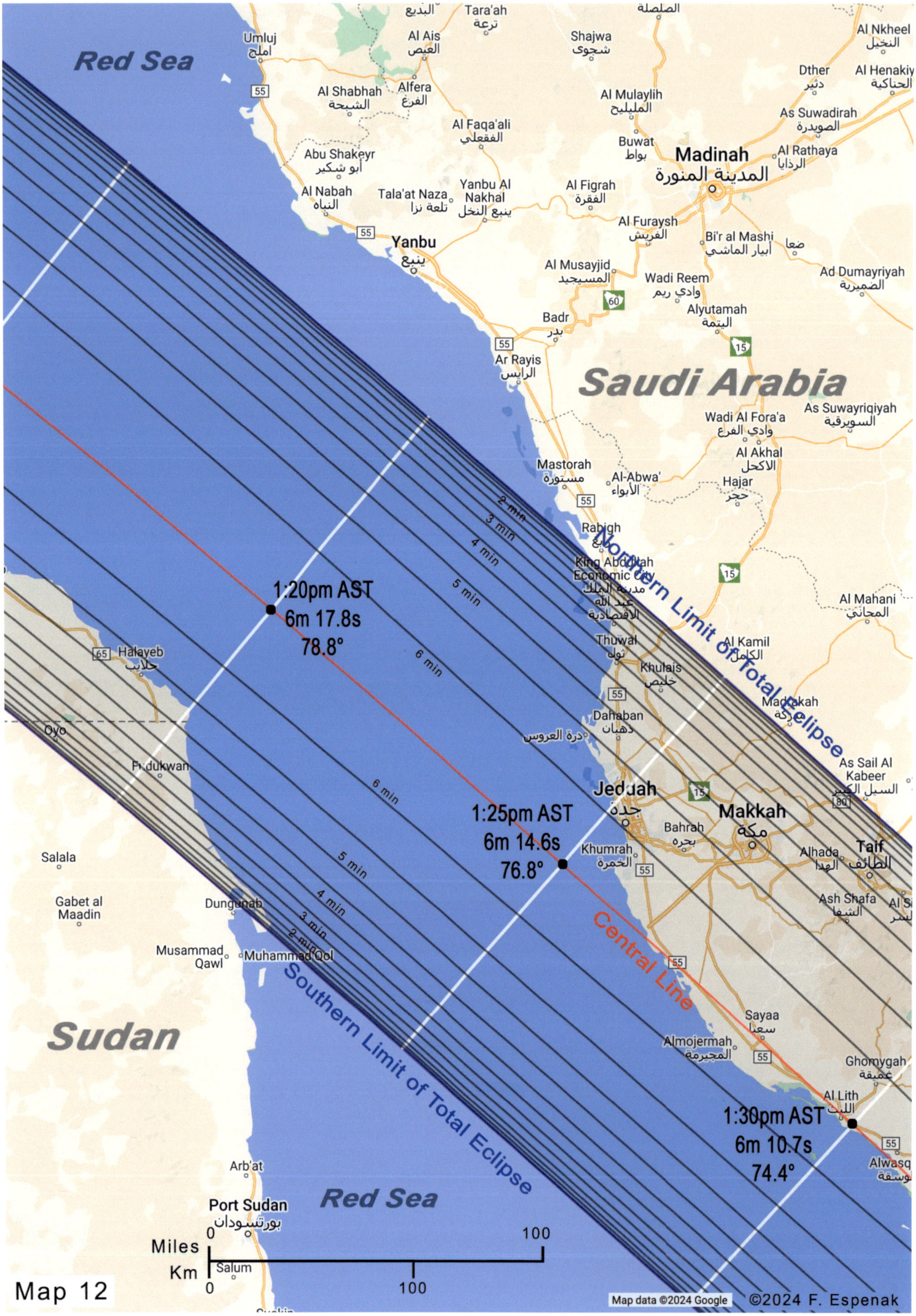
Red Sea
Saudi Arabia
Sudan
Madinah
Yanbu
Jeddah
Makkah
Taif
Port Sudan
Northern Limit of Total Eclipse
Southern Limit of Total Eclipse
Central Line
1:20pm AST
6m 17.8s
78.8°
1:25pm AST
6m 14.6s
76.8°
1:30pm AST
6m 10.7s
74.4°
2 min
3 min
4 min
5 min
6 min
Miles
Km
0
100
Map 12
Map data ©2024 Google
©2024 F. Espenak

King Abdullah Economic City
Dhubaya-Jumah
Al Farei
Al Mahani
Daghbag
Miles
Km
0
50
0
50
Al Kamil
Thuwal
Khulais
Rahat
Saudi Arabia
Al Adl
Ghran
Madrakah
Al Barzah
Dahaban
Asfan
Haddat ash Sham
Northern Limit of Total Eclipse
Al Qufayf
Al Faysaliyah
Ateef
Ashayrah
Al Mubarak
Al Jumum
Al Ju'ranah
As Sayl as Saghir
Jeddah
جدة
Makkah
مكة
Khumrah
Bahrah
Ash Shishah
Alhada
Ach Shuwaybit
As Sudayrah
Kulaikh
Taif
الطائف
Al Sir
1:25pm AST
6m 14.6s
76.8°
Malakan
2 min
3 min
4 min
5 min
5 min 30 sec
6 min
Shoqsan
Ash Shafa
Qia
Bani Saad
Central Line
Ewaha
Yalamlam
Tafel
Alsadyah
Sa'diya
bny yazed
Maysaan
Rbooa Alaean
Hadaad Bani Male
Sayaa
Almojermah
Heqal
Adham
Alqare
Red Sea
6 min
'Ushara
Alqalh
Ghomygah
Alhosainiah
Hajral
1:30pm AST
6m 10.8s
74.4°
Lith
5 min
4 min
3 min
2 min
Alwasqah
Alshwaaq
Salm Alzwaher
Hefar
doga
Almuzayl
Map 12A
Map data ©2024 Google
©2024 F. Espenak

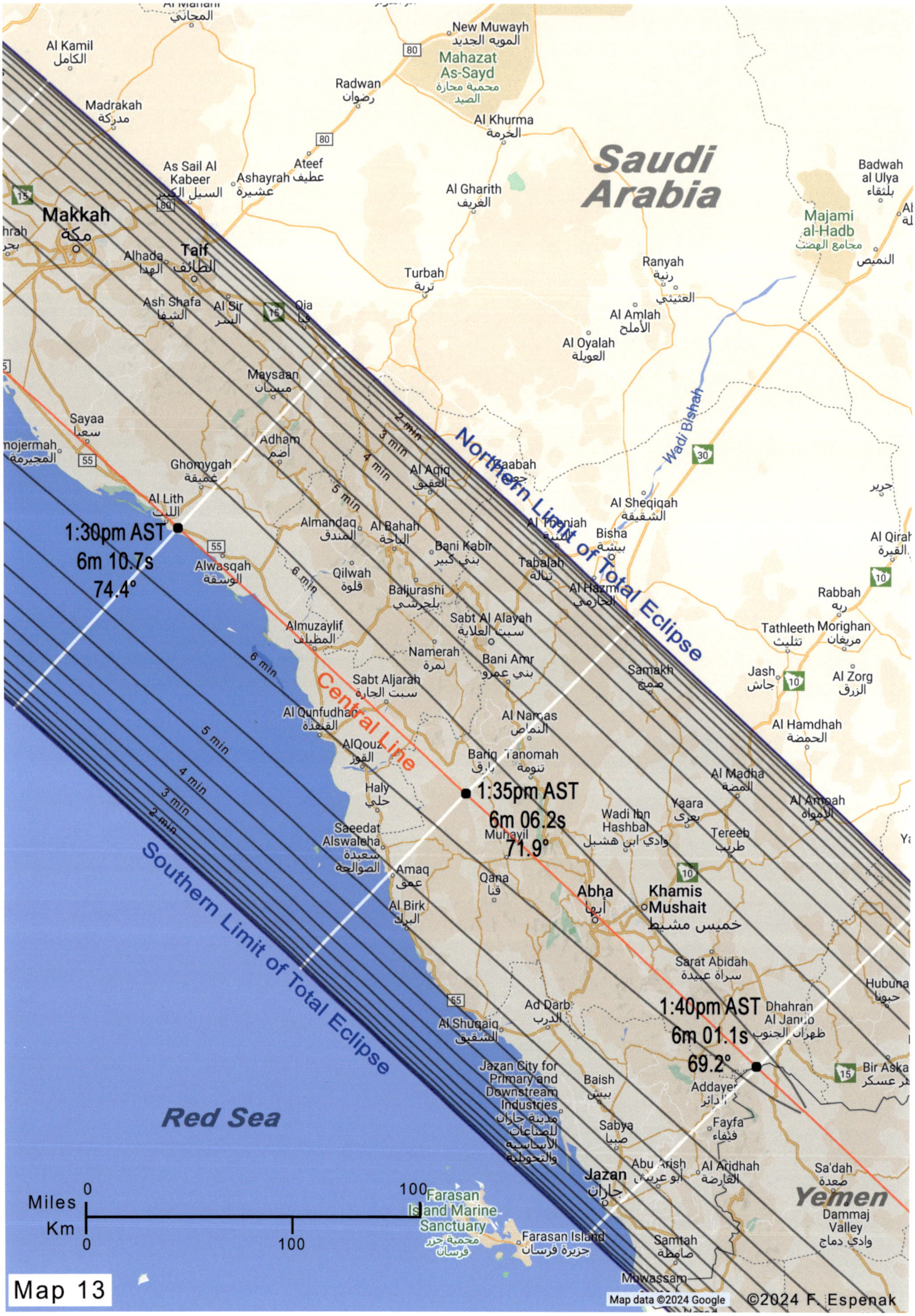
Saudi Arabia
Red Sea
Yemen
Makkah
Taif
Northern Limit of Total Eclipse
Southern Limit of Total Eclipse
Central Line
1:30pm AST
6m 10.7s
74.4°
1:35pm AST
6m 06.2s
71.9°
1:40pm AST
6m 01.1s
69.2°
2 min
3 min
4 min
5 min
6 min
6 min
5 min
4 min
3 min
2 min
Miles 0 100
Km 0 100
Map 13
Map data ©2024 Google
©2024 F. Espenak
Al Kamil
Madrakah
As Sail Al Kabeer
Ashayrah
Ateef
Radwan
New Muwayh
Mahazat As-Sayd
Al Khurma
Al Gharith
Badwah al Ulya
Majami al-Hadb
Alhada
Ash Shafa
Al Sir
Qia
Turbah
Ranyah
Al Amlah
Al Oyalah
Maysaan
Sayaa
Adham
Ghomygah
Al Lith
Almandaq
Al Bahah
Al Aqiq
Bani Kabir
Qilwah
Baljurashi
Alwasqah
Almuzaylif
Sabt Al Alayah
Namerah
Bani Amr
Sabt Aljarah
Al Qunfudhah
AlQouz
Al Namas
Bariq
Tanomah
Haly
Muhayil
Saeedat Alswaleha
Amaq
Qana
Al Birk
Abha
Khamis Mushait
Wadi Ibn Hashbal
Yaara
Tereeb
Al Madha
Al Amwah
Sarat Abidah
Dhahran Al Janub
Ad Darb
Al Shuqaiq
Jazan City for Primary and Downstream Industries
Baish
Addayer
Fayfa
Sabya
Abu Arish
Al Aridhah
Jazan
Samtah
Muwassam
Farasan Island
Farasan Island Marine Sanctuary
Sa'dah
Dammaj Valley
Bir Askar
Hubuna
Al Hamdhah
Al Zorg
Jash
Tathleeth
Morighan
Rabbah
Al Qirah
Al Sheqiqah
Bisha
Tabalah
Al Hazmi
Al Thoniah
Samakh
Wadi Bishah
Kaabah

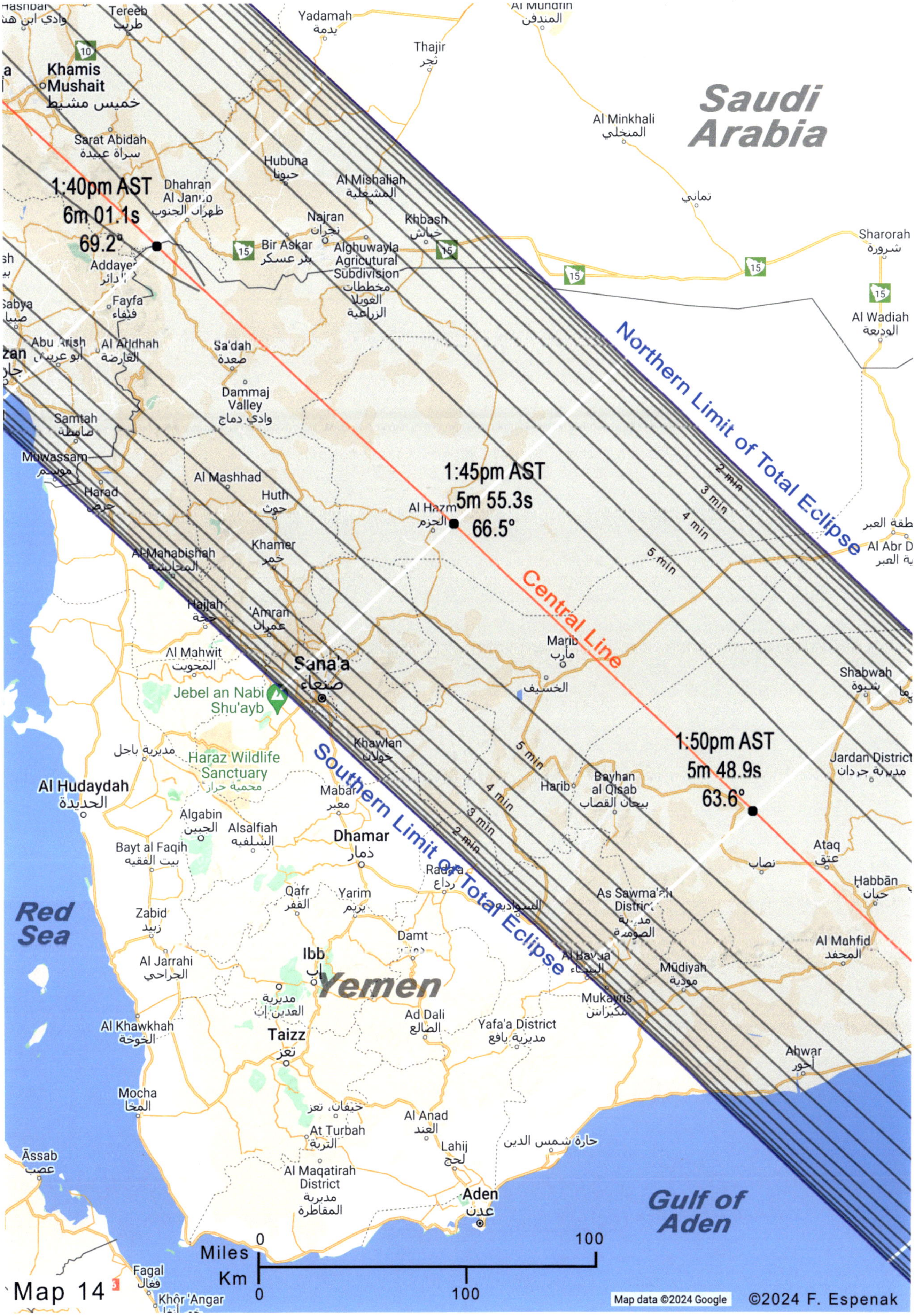
Saudi Arabia
Yemen
Red Sea
Gulf of Aden
Northern Limit of Total Eclipse
Southern Limit of Total Eclipse
Central Line
1:40pm AST
6m 01.1s
69.2°
1:45pm AST
5m 55.3s
66.5°
1:50pm AST
5m 48.9s
63.6°
2 min
3 min
4 min
5 min
Khamis Mushait
Sarat Abidah
Dhahran Al Janub
Najran
Bir Askar
Hubuna
Al Mishaliah
Khbash
Alghuwayla Agricultural Subdivision
Yadamah
Thajir
Al Minkhali
Sharorah
Al Wadiah
Addayer
Fayfa
Abu Arish
Al Aridhah
Sa'dah
Dammaj Valley
Samtah
Muwassam
Harad
Al Mashhad
Huth
Al Hazm
Khamer
Al Mahabishah
Hajjah
'Amran
Al Mahwit
Sana'a
Jebel an Nabi Shu'ayb
Haraz Wildlife Sanctuary
Khawlan
Marib
Harib
Bayhan al Qisab
Shabwah
Jardan District
Ataq
Habbān
Al Mahfid
Al Hudaydah
Algabin
Alsalfiah
Bayt al Faqih
Mabar
Dhamar
Rada'a
Qafr
Yarim
Zabid
Damt
Ibb
Al Jarrahi
As Sawma'ah District
Al Bayda
Mūdiyah
Mukayris
Ad Dali
Yafa'a District
Al Khawkhah
Taizz
Mocha
At Turbah
Al Anad
Lahij
Al Maqatirah District
Aden
Ahwar
Āssab
Fagal
Khōr 'Angar
Miles
Km
0
100
Map data ©2024 Google
©2024 F. Espenak
Map 14

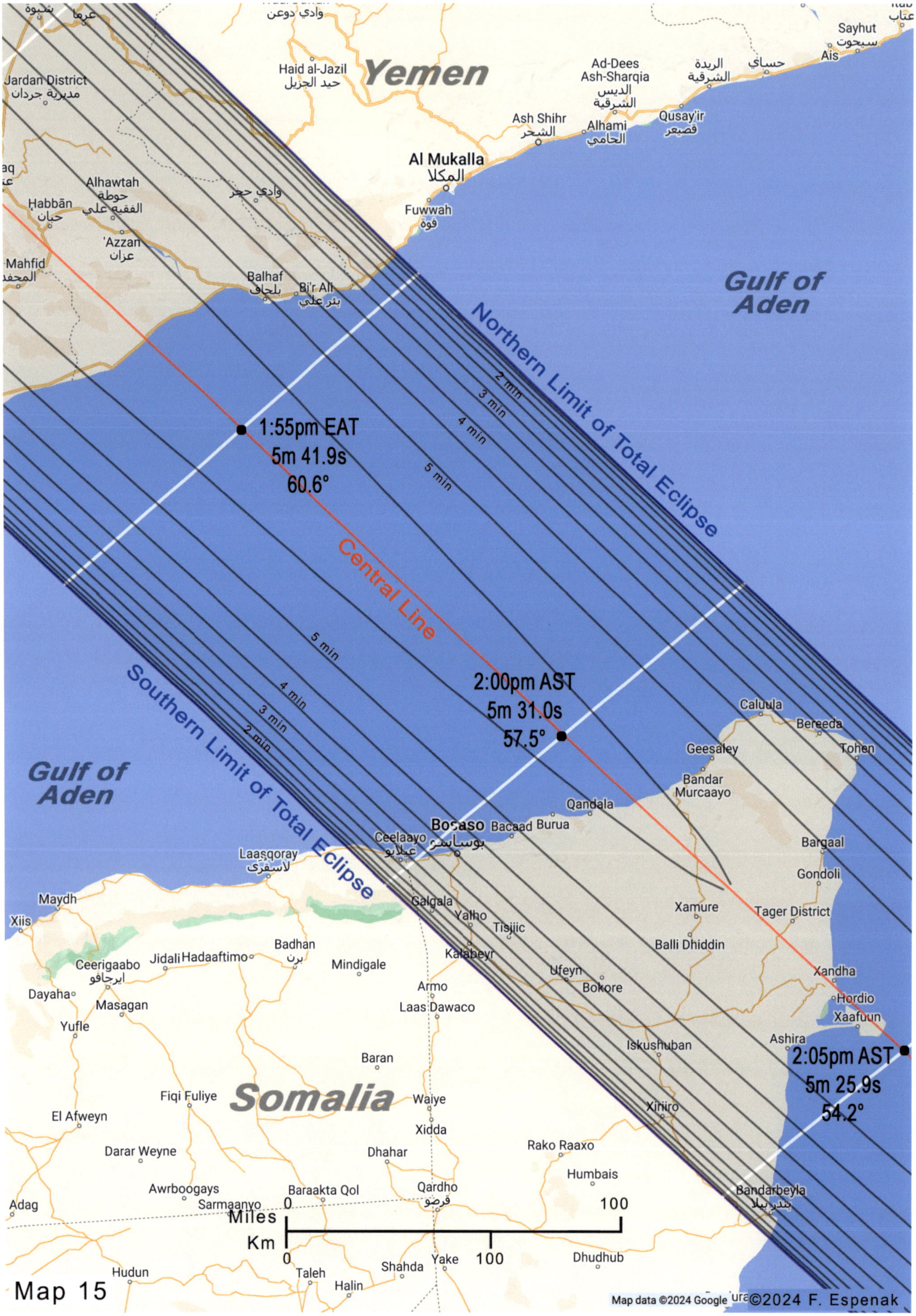
Yemen
Gulf of Aden
Gulf of Aden
Somalia
Northern Limit of Total Eclipse
Southern Limit of Total Eclipse
Central Line
2 min
3 min
4 min
5 min
5 min
4 min
3 min
2 min
1:55pm EAT
5m 41.9s
60.6°
2:00pm AST
5m 31.0s
57.5°
2:05pm AST
5m 25.9s
54.2°
Jardan District
Haid al-Jazil
Ad-Dees Ash-Sharqia
Sayhut
Ais
Ash Shihr
Alhami
Qusay'ir
Al Mukalla
Fuwwah
Alhawtah
Habbān
'Azzan
Mahfid
Balhaf
Bi'r Ali
Caluula
Bereeda
Tohen
Geesaley
Bandar Murcaayo
Qandala
Bosaso
Bacaad
Burua
Ceelaayo
Laasqoray
Bargaal
Gondoli
Maydh
Xiis
Galgala
Yalho
Tisjiic
Xamure
Tager District
Balli Dhiddin
Ceerigaabo
Jidali
Hadaaftimo
Badhan
Mindigale
Kalabeyr
Ufeyn
Bokore
Xandha
Hordio
Xaafuun
Dayaha
Masagan
Armo
Laas Dawaco
Yufle
Iskushuban
Ashira
Baran
Fiqi Fuliye
Waiye
El Afweyn
Xidda
Xiriiro
Darar Weyne
Dhahar
Rako Raaxo
Humbais
Awrboogays
Baraakta Qol
Qardho
Bandarbeyla
Adag
Sarmaanyo
Miles
Km
0
100
0
100
Hudun
Taleh
Shahda
Yake
Dhudhub
Halin
Map 15
Map data ©2024 Google
©2024 F. Espenak

Eclipse Guides from AstroPixels Publishing

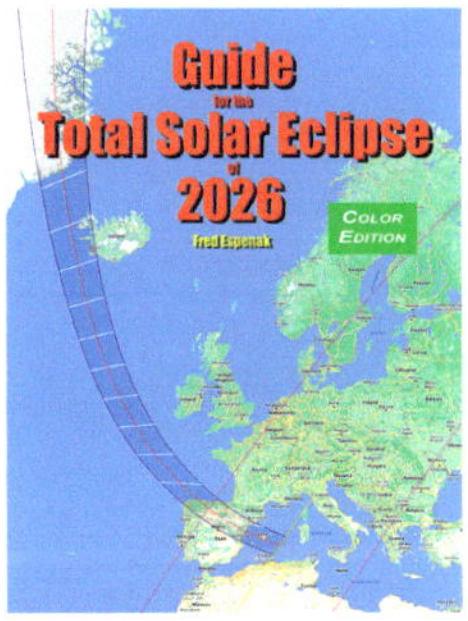

Guide for the Total Solar Eclipse of 2026

The total solar eclipse of August 12, 2026 will be visible from a path across northern Russia, the Arctic, Greenland, Iceland, and Spain. The **Guide for the Total Solar Eclipse of 2026** contains 14 full-page maps of the path. The large scale shows roads, towns and cities, rivers and lakes.

Tables for hundreds of cities provide times each eclipse phase of the along with the eclipse magnitude, duration, and the Sun's altitude. A climatology study identifies areas along the eclipse path where the highest probability of favorable weather may be found.

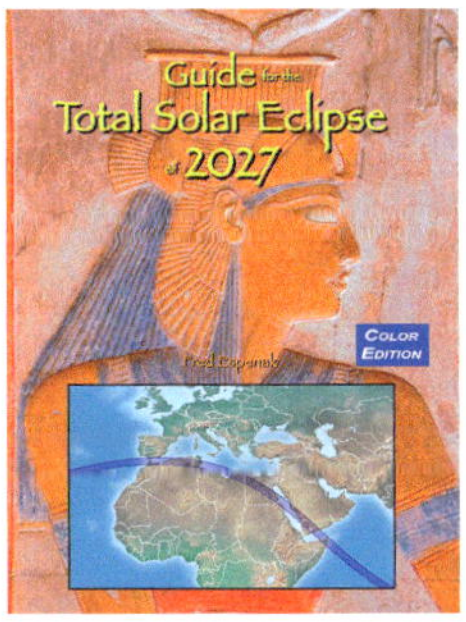

Guide for the Total Solar Eclipse of 2027

The total solar eclipse of August 2, 2027 will be visible from a path across Spain, Morocco, Algeria, Tunisia, Libya, Egypt, Sudan, Saudia Arabia, Yemen, and Somalia. The **Guide for the Total Solar Eclipse of 2027** contains 20 full-page maps of the path.

Tables for hundreds of cities provide times for each eclipse phase along with the eclipse magnitude, duration and the Sun's altitude. Cloud statistics identify areas where the highest probability of clear sky may be found.

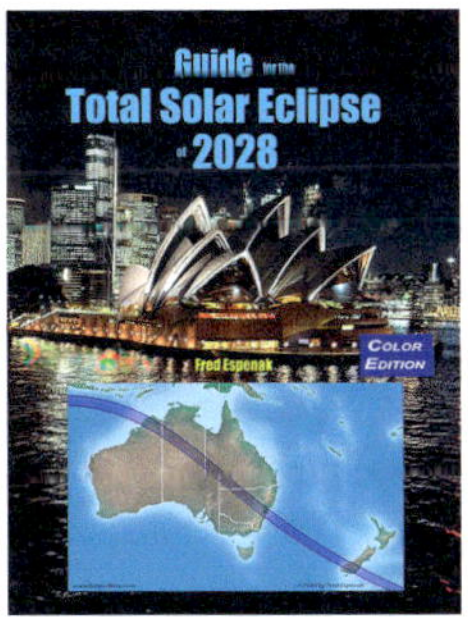

Guide for the Total Solar Eclipse of 2028

The total solar eclipse of July 22, 2028 will be visible from a path across Australia and New Zealand. The **Guide for the Total Solar Eclipse of 2028** contains 19 full-page maps of the path of totality.

Tables for over one hundred cities provide times for each eclipse phase with the duration of totality and the Sun's altitude. Cloud statistics identify areas where the highest probability of clear sky may be found. A preview is given for total eclipses through 2039.

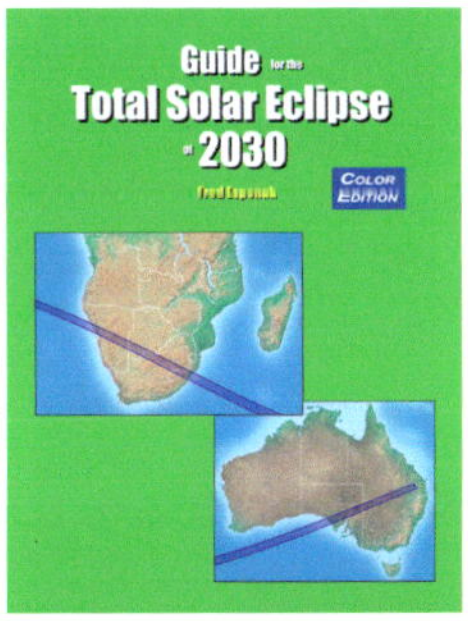

Guide for the Total Solar Eclipse of 2030

The total solar eclipse of November 25, 2030 will be visible from a path across Africa and Australia. The **Guide for the Total Solar Eclipse of 2028** contains 17 full-page maps of the path of totality.

Tables for over one hundred cities provide times for each eclipse phase with the duration of totality and the Sun's altitude. Cloud statistics identify areas where the highest probability of clear sky may be found. A preview is given for total eclipses through 2039.

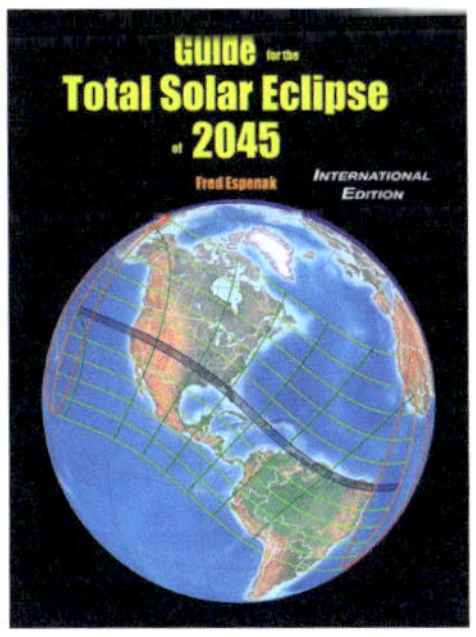

Guide for the Total Solar Eclipse of 2045

The total solar eclipse of August 12, 2045 will be visible from a path across the USA, the Caribbean, and South America. The **Guide for the Total Solar Eclipse of 2045** contains 37 full-page maps of the path of totality.

Tables for hundreds cities provide times for each eclipse phase with the duration of totality, eclipse magnitude, and the Sun's altitude. Cloud statistics identify areas where the highest probability of clear sky may be found.

For more information on each *Guide*, including sample pages of each, visit:

astropixels.com/pubs/guides.html

The *Guides* may be ordered from ***Amazon***

Eclipse Books from AstroPixels Publishing

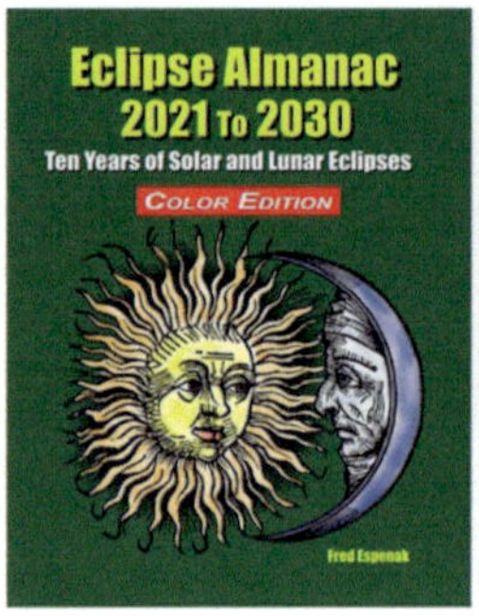

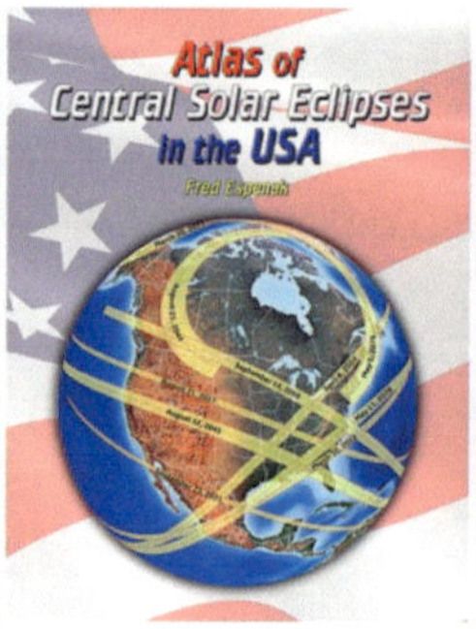

Eclipse Almanac - 2021 to 2030

A concise reference for every eclipse of the Sun and the Moon over a 10-year period. It identifies when and where each event will be seen with maps for each. Additional details about each eclipse are included. Other volumes in this series include the decades 2032 to 2040, 2041 to 2050, 2051 to 2060, and 2061 to 2070.

21st Century Canon of Solar Eclipses

The complete guide to every solar eclipse occurring from 2001 tom 2100 (224 eclipses in all). It includes information and maps for all total, annular, hybrid, and partial eclipses. A special world atlas shows detailed, full page maps of all central eclipse paths (total, annular, and hybrid).

Thousand Year Canon of Solar Eclipses — 1501 to 2500

Contains maps and data for each of the 2,389 solar eclipses occurring over the ten-century period centered on the present era. The maps are arranged twelve to a page at an image scale allowing the determination of eclipse visibility from any location on Earth. A comprehensive catalog lists the essential characteristics of each eclipse.

Atlas of Central Solar Eclipses in the USA

When was the last total eclipse through the USA and when is the next? How often do they happen? What total eclipse tracks passed across the USA during the 17th, 18th, and 19th centuries, etc., and what states did they include? And how often is a total solar eclipse visible from each of the 50 states? The Atlas of Central Solar Eclipses in the USA answers all these questions and more, with hundreds of maps and tables.

21st Century Canon of Lunar Eclipses

The complete guide to every lunar eclipse occurring from 2001 tom 2100 (228 eclipses in all). It includes information and maps for all total, partial, and penumbral eclipses. The predictions use a new model for Earth's elliptical shadows.

Thousand Year Canon of Lunar Eclipses — 1501 to 2500

Contains maps and data for each of the 2,424 lunar eclipses occurring over the ten-century period centered on the present era. The maps are arranged twelve to a page at an image scale allowing the determination of eclipse visibility from any location on Earth. A comprehensive catalog lists the essential characteristics of each eclipse.

All books are available in two editions: 1) Black and White, and 2) Color.

For more information including sample pages of each, visit:

astropixels.com/pubs/guides.html

Each *Book* may be ordered from ***Amazon***

Made in the USA
Monee, IL
12 September 2026

62974506R00029